**INTERNATIONALER DESIGNPREIS
BADEN-WÜRTTEMBERG
UND MIA SEEGER PREIS 2022**

**BADEN-WÜRTTEMBERG
INTERNATIONAL DESIGN AWARD
AND MIA SEEGER PRIZE 2022**

↗ DESIGN CENTER
BADEN-WÜRTTEMBERG

avedition

INTERNATIONALER DESIGNPREIS
BADEN-WÜRTTEMBERG
UND MIA SEEGER PREIS 2022

BADEN-WÜRTTEMBERG
INTERNATIONAL DESIGN AWARD
AND MIA SEEGER PRIZE 2022

DESIGN CENTER
BADEN-WÜRTTEMBERG

avedition

FOCUS OPEN 2022

INHALT

INHALT		2
VORWORTE		4–11
Nachhaltigkeit und Klimaschutz		
→ Dr. Patrick Rapp MdL		4
Nachhaltigkeit treibt Innovation		
→ Susanne Bay und Christiane Nicolaus		6
DIE JURY		
Roland de Fries		52
Dina Gallo		74
Joa Herrenknecht		96
Andreas Hess		112
Marc-Gregor Weidt		138
Irmy Wilms-Haverkamp		172
AUSGEZEICHNETE PRODUKTE		14–195
1	Investitionsgüter, Werkzeuge	14
2	Healthcare	48
3	Bad, Wellness	54
4	Küche, Haushalt, Tischkultur	58
5	Interior	76
6	Lifestyle, Accessoires	90
7	Licht	98
8	Consumerelectronic, Entertainment	108
9	Freizeit, Sport, Spielen	114
10	Gebäudetechnik	132
11	Public Design, Urban Design	140
12	Mobility	160
13	Service Design	174
14	Materials & Surfaces	184
INTERVIEWS		
Peter Staudinger, Rosenbauer International AG		20
Mario Zeppetzauer, Formquadrat GmbH		26
Felix Fuchs, Instagrid GmbH		32
Markus Hain und Holger De Boer, Miele & Cie. KG		64
Henk Kosche, ERCO GmbH		104
Stefan Degn, Formquadrat GmbH		120
Max Maier, Maxmaier Urbandevelopment		156
Stefan Berroth,		
Recaro Aircraft Seating GmbH & Co. KG		166
Claudia S. Friedrich, Zweigrad Design		180
Alexander Schlag, Yellow Design GmbH		190
MIA SEEGER PREIS 2022		196–213
APPENDIX A–Z		214–219
Adressen		215
Namensregister		218
Das Design Center		
→ Let's Thank		220
→ Alle Formate und Services		222
Impressum		224

CONTENTS

CONTENTS		2
FOREWORDS		4–11
Sustainability and climate protection		
→ Dr. Patrick Rapp MdL		5
Sustainability is driving innovation		
→ Susanne Bay and Christiane Nicolaus		9
THE JURY		
Roland de Fries		52
Dina Gallo		74
Joa Herrenknecht		96
Andreas Hess		112
Marc-Gregor Weidt		138
Irmy Wilms-Haverkamp		172
THE AWARD-WINNING PRODUCTS		14–195
1	Capital goods, tools	14
2	Healthcare	48
3	Bathroom, wellness	54
4	Kitchen, household, table	58
5	Interiors	76
6	Lifestyle, accessories	90
7	Lighting	98
8	Consumer electronics, entertainment	108
9	Leisure, sports, play	114
10	Building technology	132
11	Public design, urban design	140
12	Mobility	160
13	Service design	174
14	Materials & surfaces	184
INTERVIEWS		
Peter Staudinger, Rosenbauer International AG		20
Mario Zeppetzauer, Formquadrat GmbH		26
Felix Fuchs, Instagrid GmbH		32
Markus Hain and Holger De Boer, Miele & Cie. KG		64
Henk Kosche, ERCO GmbH		104
Stefan Degn, Formquadrat GmbH		120
Max Maier, Maxmaier Urbandevelopment		156
Stefan Berroth,		
Recaro Aircraft Seating GmbH & Co. KG		166
Claudia S. Friedrich, Zweigrad Design		180
Alexander Schlag, Yellow Design GmbH		190
MIA SEEGER PRIZE 2022		196–213
APPENDIX A–Z		214–219
Addresses		215
Index of names		218
The Design Center		
→ Let's thank		220
→ All formats and services		222
Publishing details		224

NACHHALTIGKEIT UND KLIMASCHUTZ

DR. PATRICK RAPP MDL

Sehr geehrte Damen und Herren,
liebe Preisträgerinnen und Preisträger,

die Designwirtschaft Baden-Württembergs ist ein wichtiger Erfolgs- und Wirtschaftsfaktor für Baden-Württemberg. Dafür spricht ein beeindruckender Gesamtumsatz in Höhe von 2,4 Mrd. Euro und das im ersten Jahr der Pandemie. Im Vor-Corona-Zeitraum war der Gesamtumsatz nochmals zehn Prozent höher. Die Prognosen für die kommenden Jahre sind positiv und lassen auf eine rasche Markterholung hoffen. Besonders stark und widerstandsfähig zeigten sich übrigens die Wirtschaftszweige Grafik- und Kommunikation, Interior und Raumgestaltung sowie der Bereich des Industrie-, Produkt- und Mode-Designs, der sich auch bei der diesjährigen Preisverleihung wieder stark präsentiert.

Im Rahmen des diesjährigen Internationalen Designpreis Baden-Württemberg Focus Open hat die Jury 44 Auszeichnungen in unterschiedlichen Kategorien vergeben – und das Niveau hat sich im Vergleich zum Vorjahr nochmals gesteigert. Beim Design setzen sich Unternehmen zunehmend auch mit den Themen Nachhaltigkeit und Klimaschutz auseinander.

Mit dem Focus Meta vergab die Jury 2022 zum dritten Mal einen Sonderpreis, der beispielhafte Lösungen für übergreifende und aktuelle Themen belohnt. Er ist ein hervorragendes und gelungenes Beispiel dafür, dass Unternehmen verstärkt nachhaltig denken – und Initiative ergreifen. Dass er in diesem Jahr für ein Projekt aus »The LÄND«, genauer aus Ludwigsburg, vergeben wird, freut uns natürlich besonders. Das Projekt integriert aktuellste technische Optionen für eine klimaneutrale und nachhaltige Immobiliennutzung. Neugierig geworden?

Das aktuelle Jahrbuch liefert Ihnen wieder Hintergrundinformationen zu den Entstehungsprozessen der Projekte und die konkrete Arbeit von Designerinnen und Designern. Sie erfahren beispielsweise mehr über einen Flugzeugsitz, der aufgrund seiner besonderen Bauweise einen voll ausgelasteten Jet um rund 420 Kilogramm leichter werden lässt und somit die CO_2-Emission pro Platz und Jahr um 108 Kilogramm reduziert. Oder einen Feuerwehrhelm, der nur 1,23 Kilogramm wiegt und von der passenden Helmlampe und einer Wärmebildkamera bis hin zur Atemschutzmaske und einer rot leuchtenden Positionsleuchte alles enthält. Ein T-Shirt für Elektroinstallateure mit integrierter Notruffunktion, das notfalls nach Stromschlägen ein Signal an die nächste Notrufzentrale absetzen kann, rettet nicht nur Leben, sondern sieht auch noch schick aus.

Diese Beispiele machen bereits den Wert durchdachter und nutzerorientierter Gestaltung deutlich. Designerinnen und Designer sind Impulsgeber und Innovationstreiber. Sie unterstützen Unternehmen im weltweiten Wettbewerb und tragen durch ihr Know-how nicht selten zu Alleinstellungsmerkmalen und zum Wettbewerbsvorsprung bei.

Ein herzliches Dankeschön gilt daher allen Interviewpartnerinnen und -partnern, die uns an ihrem Know-how teilhaben lassen, sowie allen, die sich in diesem Jahr beim Focus Open dem internationalen Wettbewerb gestellt haben. Bedanken will ich mich auch bei den Mitarbeiterinnen und Mitarbeitern des Design Center Baden-Württemberg und der Jury dafür, dass auch 2022 die Erfolgsgeschichte des Focus Open fortgeschrieben werden konnte.

Im Namen der Landesregierung von Baden-Württemberg wünsche ich weiterhin viel Erfolg und Mut zu neuen Perspektiven, innovativen Produkten sowie kreativen und nachhaltigen Designs, die Freude machen und Mensch und Umwelt Nutzen bringen.

DR. PATRICK RAPP MDL
Staatssekretär für Wirtschaft,
Arbeit und Tourismus des
Landes Baden-Württemberg

SUSTAINABILITY AND CLIMATE PROTECTION

DR. PATRICK RAPP MDL

Dear readers,
Dear prize winners,

Baden-Württemberg's design sector is an important success factor for the state and plays a valuable role in its economy – as is apparent from its impressive total turnover of €2.4 billion in 2020, despite the fact that we were experiencing the first year of the pandemic. Before corona, total turnover was at an even higher level (+10%). The forecasts for the coming years are positive and suggest the market will recover quickly. Segments that proved particularly strong and resilient included graphics and communication design, interior design, as well as industrial, product and fashion design, all of which are once again well represented among this year's award winners.

In the course of this year's Baden-Württemberg International Design Award Focus Open, the jury presented 44 accolades in various categories – and the calibre is even higher than last year. When it comes to design, companies are increasingly tackling the issues of sustainability and climate protection as well.

In 2022, the jury presented a Focus Meta award for the third time – a special prize for exemplary solutions that address overarching topical issues. It is an outstanding and successful example of how companies are thinking more and more sustainably – and taking the initiative. We're particularly delighted that this year's accolade went to an entry from Baden-Württemberg – or »The LÄND« as we call it: a project from Ludwigsburg that integrates state-of-the-art technical options for the climate-neutral and sustainable use of real estate. Curious?

As always, the current edition of the yearbook provides background information on the processes behind the projects and the concrete work of the designers involved. As you turn the pages, you'll learn more about products such as an aircraft seat that reduces the weight of a fully occupied jet by approx. 420 kilograms thanks to its special construction – thereby reducing CO_2 emissions by 108 kilograms per seat and year. Or a firefighting helmet that weighs just 1.23 kilograms and can be equipped with everything from a helmet light and thermal imaging camera all the way to a respirator mask and a position light that glows red in the dark. You'll also discover a T-shirt for electricians with an integrated emergency alert function; besides saving lives by sending a signal to the nearest emergency call centre if the wearer suffers an electric shock, it looks great as well.

These examples demonstrate the value of well thought-through and user-oriented design. As a source of fresh impetus and innovation drivers, designers provide key support for companies that have to compete in the global market: thanks to their expertise, they are able to make important contributions to unique selling propositions and enhance their clients' competitive edge.

I would therefore like to say a sincere thank you to all the interviewees who shared their expertise with us, as well as to everybody who entered this year's Focus Open to face international competition. A big thank you also to the staff of the Design Center Baden-Württemberg and the jury for ensuring that the Focus Open success story could be continued in 2022.

On behalf of the state government of Baden-Württemberg, I wish you every success for the future and the courage to embrace new perspectives, innovative products and creative, sustainable designs that not only give pleasure but benefit both people and the environment.

DR. PATRICK RAPP MDL
State Secretary of Economic Affairs, Labour and Tourism Baden-Württemberg

FOCUS OPEN 2022
NACHHALTIGKEIT TREIBT INNOVATION

KLIMAGERECHTES DESIGN ERÖFFNET CHANCEN

Das Thema Nachhaltigkeit ist in aller Munde. Leider wird es inzwischen fast inflationär eingesetzt und nicht selten als Feigenblatt für anscheinend klimafreundliche Lösungen verwendet. Oft allein auf Material oder Verfahren reduziert, fällt die Auseinandersetzung mit den Bestimmungsgrößen für Nachhaltigkeit in ihrer Gänze meist viel zu kurz und zu oberflächlich aus. Zugegeben: Das Themenfeld ist riesig, vielschichtig und zu großen Teilen mit Unbekanntem und Unvorhersehbarem behaftet. Es fällt schwer, es in seiner Vollständigkeit zu erfassen. Das lässt zögern.

Dennoch: Wenn wir das globale Klima vor dem Kollaps bewahren wollen, müssen wir alle unseren Beitrag dazu leisten. Und hier kommt Designleistung ins Spiel. Designer:innen haben großen Einfluss auf unsere Produktwelt – und damit eine große Verantwortung. Oder genauer gesagt, sie können den ökologischen Fußabdruck eines Produktes stark beeinflussen, positiv wie negativ. Bis zu 80 Prozent der Umweltauswirkungen eines Produktes werden durch dessen Design vorbestimmt, so hat es das Bundesministerium für Umwelt, Naturschutz, nukleare Sicherheit und Verbraucherschutz beziffert.

RETHINK:DESIGN, DAS NEUE FORMAT DES DESIGN CENTER BADEN-WÜRTTEMBERG

Aus diesem Grund haben wir Anfang des Jahres RETHINK:DESIGN gestartet, eine thematisch auf den Klima-Impact des Designs fokussierte Reihe mit Best-Practice-Beispielen, Interviews, Vorträgen und Workshops. Damit wollen wir dazu animieren, eingefahrene Prozesse zu hinterfragen und Nachhaltigkeit als Innovationstreiber zu begreifen. Designer:innen wollen wir darin bestärken, sich dem Klimaschutz noch konsequenter anzunehmen. Entscheiderinnen und Entscheidern möchten wir zeigen, wie man diese erweiterte Designkompetenz gekonnt einsetzt. Letztlich geht es darum, dass soziales und nachhaltiges Handeln eine gute Basis für wirtschaftlichen Erfolg darstellt.

DIE RELEVANZ ERKENNEN

Viele Unternehmen haben die Relevanz des Klimathemas durchaus erkannt, zögern aber noch bei der Transformation. Auch, weil sie nicht wissen, wie und wo sie dieses umfangreiche Thema anpacken sollen. Denn konsequent integrierte Nachhaltigkeit bedeutet das Hinterfragen und Ändern von Prozessen, das Erfüllen von Auflagen, das lückenlose Dokumentieren, etwa in Form von Nachhaltigkeitsberichten und vieles mehr. Monetärer und zeitlicher Aufwand sind von vornherein nicht einschätzbar, das unternehmerische Risiko nicht vollständig kalkulierbar.

DEN EINSTIEG FINDEN

Unternehmer:innen zögern oft und scheuen Neues, weil sie den Aufwand und die Tragweite von Projekten oder Aktivitäten nicht von vornherein einschätzen können. Daher ist es umso wichtiger, einen Einstieg zu finden, der umsetzbar ist und eine Agenda zu entwickeln, die realistisch, durchhaltbar und erfolgversprechend erscheint.

Gerade was die Klimarelevanz von Produkten oder Dienstleistungen angeht, können Designer:innen wichtige Überzeugungsarbeit leisten, Alternativen und machbare Lösungen aufzeigen. Schließlich ist Design, wenn konsequent im Unternehmen implementiert, eine der wenigen Disziplinen, die in Entwicklungsprozessen durchgängig präsent sind, von der ersten Idee bis zur Markteinführung – und darüber hinaus. Designer:innen können also großen Einfluss nehmen, auch auf die Klimaverträglichkeit neuer Konzepte. So wie sie schon heute im Bereich Innovationsmanagement involviert sind, können sie Unternehmen durchaus auch in Klimafragen beraten. Beispielsweise in interdisziplinären CSR-Teams (Corporate Social Responsibility). Das verlangt natürlich erweiterte Kompetenzen, was für die Designbranche aber durchaus eine Investition in neue Geschäftsbereiche sein kann.

DESIGN EINBINDEN

Wie schon erwähnt: Design hat einen großen Anteil an den Umweltauswirkungen eines Produktes. Das ist allerdings keine revolutionäre Erkenntnis, denn die Berücksichtigung ökologischer Aspekte ist für die Designbranche ohnehin selbstverständlich. Man muss das Rad also gar nicht neu erfinden, sondern – bildlich gesprochen – ein paar Speichen hinzufügen

SUSANNE BAY
Regierungspräsidentin
Regierungsbezirk Stuttgart

und es konsequenter drehen. Unsere Botschaft an Unternehmer:innen lautet daher: Setzen Sie Ihr über viele Jahre erlangtes Wissen um und zögern Sie nicht, es jetzt in Richtung Klimaschutz zu justieren. Verlassen Sie alte Pfade und denken Sie neu – und binden Sie dazu auch Designleistung ein. Die Designbranche kann diesen Transformationsprozess anschieben, unterstützen und konkretisieren. Letztlich geht es auch darum, Begeisterung zu erzeugen – das wiederum kann kaum eine andere Berufsgruppe besser. So gesehen können Nachhaltigkeit und aktiver Klimaschutz zum neuen Innovationstreiber werden.

STELLSCHRAUBEN DREHEN

Es gibt viele Schrauben, an denen sich schon immer drehen ließ. Dazu gehören beispielsweise material- und fertigungsgerechtes Entwerfen, langlebige Formensprache, Demontagefähigkeit, Vermeidung von Verbundwerkstoffen, das Hinterfragen von Materialien, den eigenen Produktionsverfahren und denen der Zulieferer. Alle diese Stellschrauben sollten jetzt noch konsequenter hinsichtlich ihres Einflusses auf den Aspekt der Klimafreundlichkeit betrachtet werden. Das sollte für alle Entwicklungspartner in den Briefings an erster Stelle stehen. Würde man alle relevanten Umsetzungsparameter am Aspekt der Klimaschonung spiegeln, wären wir schon unglaublich weit.

Ein neues Material zum Beispiel setzt man natürlich nicht von heute auf morgen ein. Die Umstellung von Prozessen ist ein langjähriger Weg, der in Unternehmen zuweilen auf Widerstand stößt. Aber dennoch müssen wir die Dinge und Determinanten neu betrachten und konsequent neu umsetzen.

Selbstverständlich gibt es auch Stellschrauben, an denen man beispielsweise als externer Dienstleister nicht drehen kann. Daher erscheint es besonders wichtig, dass Designer:innen alle Spielräume und Möglichkeiten nutzen, die sie bekommen und insbesondere bei langjährigen Kooperationen ihre Grenzen immer wieder austesten und erweitern.

Viele gute Beispiele in diesem Buch zeigen, dass dieser Weg zum Erfolg führen kann. Wir wünschen Ihnen viel Freude beim Lesen!

VIELEN DANK!

Nun gibt es bei Wettbewerben nicht nur Gewinnerinnen und Gewinner, auch wenn lediglich diese hier im Jahrbuch gezeigt werden, viele Einreichungen haben keine Auszeichnung erhalten, die Gründe sind vielfältig und liegen in den Händen der unabhängig bewertenden Fachjury. Doch lassen Sie uns an dieser Stelle auch diesen Einreichenden sagen: Vielen Dank für Ihre Teilnahme! Und zugleich möchten wir appellieren: Bleiben Sie dran! Betrachten Sie die Nicht-Auszeichnung als Ansporn, als Optimierungsimpuls. Wagen Sie einen neuen Anlauf im nächsten Jahr – mit neuen Ideen und neuem Schwung. Sie können viel gewinnen und nur wenig verlieren – der Focus Open schont Ihre Etats, weil die Einreichungsgebühren bewusst gering sind. Damit können auch kleinere Unternehmen und Start-Ups problemlos teilnehmen: Der Focus Open ist somit auch ein Förderpreis. Eine Auszeichnung mit dem Focus Open hat sich schon für so manches neue Unternehmen als Booster erwiesen.

Danken möchten wir natürlich auch sehr herzlich unseren diesjährigen Jury-Mitgliedern. Alle Einreichungen wurden gründlich begutachtet und ausführlich diskutiert. Danke in diesem Zusammenhang auch für die vielschichtigen, konstruktiven und lebendigen Gespräche – und teils auch sehr interessanten Grundsatzdiskussionen – rund um die eingereichten, sehr unterschiedlichen Designlösungen. Die Jurytage waren für uns alle eine wertvolle Bereicherung!

HERZLICHEN GLÜCKWUNSCH!

Wir beglückwünschen die diesjährigen Preisträgerinnen und Preisträger zu ihren Auszeichnungen. Wir wünschen ihnen für ihre innovativen Produkt- und Konzeptlösungen viel Erfolg sowie weiterhin wertvolle und erfolgreiche Kooperationen – deren Ergebnisse hoffentlich wieder den Weg zum Focus Open finden werden. Dann mit neuer Jury, aber der gewohnten Neutralität, Seriosität und Offenheit, die den Internationalen Designpreis Baden-Württemberg auszeichnen.

CHRISTIANE NICOLAUS
Direktorin
Design Center Baden-Württemberg

KATEGORIEN

1. INVESTITIONSGÜTER, WERKZEUGE
2. HEALTHCARE
3. BAD, WELLNESS
4. KÜCHE, HAUSHALT, TISCHKULTUR
5. INTERIOR
6. LIFESTYLE, ACCESSOIRES
7. LICHT
8. CONSUMERELECTRONIC, ENTERTAINMENT
9. FREIZEIT, SPORT, SPIELEN
10. GEBÄUDETECHNIK
11. PUBLIC DESIGN, URBAN DESIGN
12. MOBILITY
13. SERVICE DESIGN
14. MATERIALS & SURFACES

KRITERIEN

- ✓ GESTALTUNGSQUALITÄT
- ✓ FUNKTIONALITÄT
- ✓ INNOVATIONSHÖHE
- ✓ ERGONOMIE
- ✓ INTERFACE DESIGN/ CONNECTIVITY
- ✓ USABILITY
- ✓ NACHHALTIGKEIT
- ✓ ÄSTHETIK
- ✓ BRANDING
- ✓ ENTWICKLUNGSVORSPRUNG
- ✓ USER JOURNEY
- ✓ DIGITALE INTELLIGENZ

FOCUS OPEN 2022

44 PREISTRÄGER
9 GOLD-AWARDS
18 SILVER-AWARDS
16 SPECIAL MENTION AWARDS
1 META AWARD

DIE JURY

- ✓ ROLAND DE FRIES
- ✓ DINA GALLO
- ✓ JOA HERRENKNECHT
- ✓ ANDREAS HESS
- ✓ MARC-GREGOR WEIDT
- ✓ IRMY WILMS-HAVERKAMP

CLIMATE-FRIENDLY DESIGN CREATES OPPORTUNITIES

Sustainability is a hot topic. Unfortunately, however, the term is meanwhile being used in an almost inflationary way and not infrequently as a fig leaf for solutions that are only seemingly climate-friendly. Often reduced to nothing but the materials or processes involved, the consideration given to the parameters that define sustainability in all its many facets is usually far too superficial and inadequate. Admittedly, it's a huge, multifaceted topic that can entail a great many unknowns and unforeseeable outcomes. It's difficult to grasp in its entirety. That causes hesitation.

Nevertheless: if we want to save the global climate from collapse, we all have to do our bit. And that's where design comes into play. Designers have a big impact on our product world – and that means a big responsibility too. To be more precise: they can have a major influence on a product's ecological footprint, in both a positive and a negative sense. Up to 80% of a product's environmental impact is predetermined by its design, according to Germany's Ministry of the Environment, Nature Conservation, Nuclear Safety and Consumer Protection.

RETHINK:DESIGN – THE DESIGN CENTER BADEN-WÜRTTEMBERG'S NEW FORMAT

That's why, at the beginning of this year, we launched RETHINK:DESIGN, a series featuring examples of best practice, interviews, talks and workshops that focus on design's impact on the climate. Our goal is to motivate people to question habitual processes and think of sustainability as an innovation driver. We want to encourage designers to be even more committed to addressing climate protection. We want to show decision-makers how to make clever use of this enhanced design expertise. And last but not least, we want to demonstrate that social and environmental responsibility are a good basis for commercial success.

RECOGNISING THE RELEVANCE OF THE CLIMATE ISSUE

While many companies have realised the relevance of the climate issue, they are still hesitating to embrace transformation. Partly also because they don't know how or where to begin tackling this vast topic. Because integrating sustainability systematically means questioning and changing processes, complying with requirements, end-to-end documentation, e.g. in the form of sustainability reports, and much, much more besides. At the outset, it's impossible to know how much time and money will need to be invested, and the risks involved for the business are not entirely predictable.

GETTING STARTED

Entrepreneurs and executives often hesitate and shy away from anything new because they can't judge the cost and consequences of the projects or activities involved at the outset. That's why it's all the more important to find a feasible starting point and develop an agenda that seems realistic, maintainable and promising.

Especially where the climate relevance of products or services is concerned, designers can make an important contribution by convincing people, by pointing out alternatives and viable solutions. After all, when a company implements design consistently, it's one of the few disciplines that is involved throughout the entire development process, from the initial idea all the way to launch – and beyond. That means designers can have considerable influence, including on the climate compatibility of new concepts. Just as they're already involved in the field of innovation management, they can advise companies on climate-related issues as well – for instance in interdisciplinary CSR teams (corporate social responsibility). Obviously that means having to expand their expertise, but from the design sector's perspective that can definitely be seen as an investment in new business areas.

INTEGRATING DESIGN

As already mentioned, design plays a major part in a product's impact on the environment. But that's not a revolutionary insight: taking ecological aspects into account is a matter of course for the design sector anyway. So it's not a case of having to reinvent the wheel; figuratively speaking, it's just a question of adding a few spokes and turning it more single-mindedly.

SUSANNE BAY
President,
Stuttgart District Government

CHRISTIANE NICOLAUS
Director,
Design Center Baden-Württemberg

That's why our message to entrepreneurs and businesses is this: put the knowledge you've acquired over the years into practice and don't hesitate to start gearing it towards climate protection. Leave well-trodden paths and rethink things – and integrate design to help you accomplish your goals. The design sector can kickstart this transformation process, provide support and help translate it into concrete action. At the end of the day it's about sparking enthusiasm too – and there are few professions that are better equipped to do that. If you look at it that way, sustainability and active climate protection can become a new innovation driver.

MAKING THE NECESSARY ADJUSTMENTS
There are plenty of areas where it's always been possible to make adjustments: an approach that optimises the use of materials and production processes, for instance, an enduring design language, ease of disassembly and the avoidance of composites, as well as scrutinising materials, your own production processes and those of your suppliers. Now it's a question of taking a new, more systematic look at all these areas in terms of their influence on the aspect of climate friendliness. That should be the top priority for all the development partners in the briefings. If all the relevant implementation parameters were considered from the perspective of their climate impact, we'd already have come a long way.

Obviously you don't start using a new material from one day to the next, for instance. Adapting processes is a path that takes years to complete and sometimes meets with resistance from the companies involved. But even so, we need to take a new look at things and factors, and come up with new ways to systematically translate those insights into reality.

There are some areas that e.g. an external service provider can't influence, of course. That's why it's particularly important for designers to make the most of all the scope and possibilities they have and keep testing the limits and pushing the envelope, especially in the case of long-term collaborations.

Many good examples in this book demonstrate that this path can lead to success. We hope you enjoy the read!

THANK YOU!
Competitions don't only have winners, even if they are the only participants who appear in this yearbook. Many entries did not receive an award or distinction; the reasons are many and varied, and entirely in the hands of the independent jury of experts. But we would like to take this opportunity to acknowledge those entrants too: thank you for taking part! And don't be discouraged! See the non-award as an incentive, as impetus for optimisation. Try again next year – with new ideas and new momentum. There's a lot to win and very little to lose – Focus Open is budget-friendly because the entry fees are deliberately kept low. That means even smaller companies and start-ups can easily take part: Focus Open is thus a newcomer award too, and has boosted the fortunes of many a new company over the years.

We would also like to say a heartfelt thank you to the members of this year's jury. All the entries were examined thoroughly and discussed in depth. Thank you too for the complex, constructive and lively discussions about the extremely wide-ranging entries – and for the very interesting debates about fundamental principles that they sometimes gave rise to as well. The judging was a valuable and highly enjoyable experience for all of us!

CONGRATULATIONS!
Our congratulations to this year's winners on their well-deserved awards. We wish them great success with their innovative products and concepts and hope they continue to benefit from worthwhile and productive collaborations – the results of which we'd be delighted to see entered for a future edition of Focus Open. The jury will be new, but the competition will be held with the same neutrality, integrity and open-mindedness that have always been the hallmarks of the Baden-Württemberg International Design Award.

CATEGORIES

1. CAPITAL GOODS, TOOLS
2. HEALTHCARE
3. BATHROOM, WELLNESS
4. KITCHEN, HOUSEHOLD, TABLE
5. INTERIORS
6. LIFESTYLE, ACCESSORIES
7. LIGHTING
8. CONSUMER ELECTRONICS, ENTERTAINMENT
9. LEISURE, SPORTS, PLAY
10. BUILDING TECHNOLOGY
11. PUBLIC DESIGN, URBAN DESIGN
12. MOBILITY
13. SERVICE DESIGN
14. MATERIALS & SURFACES

CRITERIA

- ✓ DESIGN QUALITY
- ✓ FUNCTIONALITY
- ✓ INNOVATIVENESS
- ✓ ERGONOMICS
- ✓ INTERFACE DESIGN/ CONNECTIVITY
- ✓ USABILITY
- ✓ SUSTAINABILITY
- ✓ AESTHETICS
- ✓ BRANDING
- ✓ STEP CHANGE IN DEVELOPMENT
- ✓ USER JOURNEY
- ✓ DIGITAL INTELLIGENCE

FOCUS OPEN 2022

- 44 PRIZE WINNERS
- 9 GOLD AWARDS
- 18 SILVER AWARDS
- 16 SPECIAL MENTION AWARDS
- 1 META AWARD

THE JURY

- ✓ ROLAND DE FRIES
- ✓ DINA GALLO
- ✓ JOA HERRENKNECHT
- ✓ ANDREAS HESS
- ✓ MARC-GREGOR WEIDT
- ✓ IRMY WILMS-HAVERKAMP

① **ROLAND DE FRIES**
Hudson Vandam LLC,
New York
hudsonvandam.com

→ **SEITE/PAGE**
52

② **DINA GALLO**
TRUMPF SE & Co. KG
(Holding), Ditzingen
trumpf.com

→ **SEITE/PAGE**
74

③ **JOA HERRENKNECHT**
Studio Joa Herrenknecht,
Berlin/Toronto
joa-herrenknecht.com

→ **SEITE/PAGE**
96

④ **ANDREAS HESS**
White ID,
Schorndorf
white-id.com

→ **SEITE/PAGE**
112

⑤ **MARC-GREGOR WEIDT**
Einmaleins GmbH,
Burgrieden
einmaleins.net

→ **SEITE/PAGE**
138

⑥ **IRMY WILMS-HAVERKAMP**
Haverkamp Interior
Design, Herford
koehlerwilms.de

→ **SEITE/PAGE**
172

1 → SEITE/PAGE
16–21

2 → SEITE/PAGE
22–27

3 → SEITE/PAGE
28–33

4 → SEITE/PAGE
34, 40

5 → SEITE/PAGE
35, 41

6 → SEITE/PAGE
36, 42

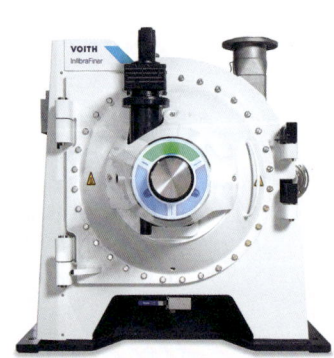

7 → SEITE/PAGE
37, 43

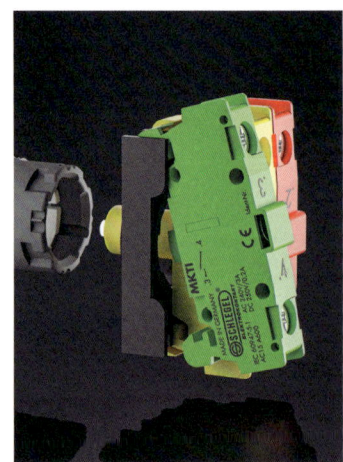

8 → SEITE/PAGE
38, 44

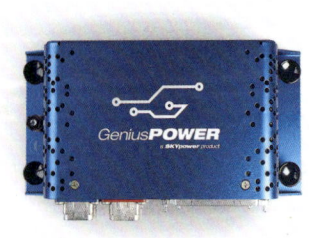

9 → SEITE/PAGE
39, 45

INVESTITIONSGÜTER, WERKZEUGE
CAPITAL GOODS, TOOLS

GOLD:
1. **HEROS H30**
   Rosenbauer International AG
   Leonding
   Österreich/Austria

2. **OCF**
   Agilox Services GmbH
   Neukirchen bei Lambach
   Österreich/Austria

3. **INSTAGRID ONEMAX**
   Instagrid GmbH
   Ludwigsburg

SILVER:
4. **DUOXPAND**
   Fischerwerke GmbH & Co. KG
   Waldachtal

5. **STE300**
   Bessey Tool GmbH & Co. KG
   Bietigheim-Bissingen

6. **FTS**
   Bleichert Automation
   GmbH & Co. KG
   Osterburken

7. **INFIBRAFINER**
   Voith Paper
   Ravensburg

8. **MK**
   Georg Schlegel GmbH & Co. KG
   Dürmentingen

SPECIAL MENTION:
9. **GENIUSPOWER**
   KID Systems GmbH
   Buxtehude

Funktionalität und Design ergänzen sich ideal – das beweisen ganz klar Maschinen oder Tools für den Profi-Bereich. Industriedesign strukturiert Bedienabläufe, optimiert die Ergonomie, treibt Innovationen voran und verbessert im Idealfall sogar die ökologische Bilanz. Und: Design hilft, die Kooperation mit autonom agierenden Anlagen auf ein menschengerechtes Level zu heben.

Functionality and design can complement each other in ideal fashion – as is vividly demonstrated by machines or tools for professional use. Industrial design structures operating procedures, optimises ergonomics, drives innovations and, at its best, even improves the ecological footprint. In addition, design helps to elevate cooperation with autonomously operating equipment to a human-friendly level.

GOLD — HEROS H30 — FEUERWEHRHELM / FIREFIGHTING HELMET

JURY STATEMENT

Dem Design ist es beispielhaft gelungen, Funktionalität und Emotionalität zu verbinden – und das bei einem primär sicherheitsrelevanten Produkt. Die Konzeption eines Profihelms ist generell eine sehr komplexe Aufgabe, weil hier viele Faktoren auf kleinstem Raum zusammenkommen. Diese Verbindung aller Anforderungen wurde perfekt gelöst.

The designers have succeeded in creating an exemplary combination of functionality and emotionality – an unusual approach for a product that is primarily safety-related. Developing a professional helmet is an extremely complex task in general because a great many factors have to be taken into account despite the limited space available. In this case, all the relevant requirements have been perfectly met.

HERSTELLER/MANUFACTURER
Rosenbauer International AG
Leonding
Österreich/Austria

DESIGN
Formquadrat GmbH
Linz
Österreich/Austria

VERTRIEB/DISTRIBUTOR
Rosenbauer Deutschland GmbH
Luckenwalde

Ein Feuerwehrhelm ist für Extremsituationen konzipiert und dient in erster Linie dem Schutz der Trägerin oder des Trägers. Gefordert ist daher eine Kombination aus größter Stabilität und Trage-Ergonomie. Der 1,23 Kilogramm leichte Heros H30 bietet daher nicht nur eine optimierte Anpassung an den Kopfumfang; seine Schale lässt sich auch so verschieben, dass der Helm nicht verrutscht, sondern stets in der Balance bleibt. Außerdem verfügt der Helm über funktionale Erweiterungsoptionen, etwa die werkzeuglose Ergänzung mit einer passenden Helmlampe, einer Wärmebildkamera oder der Atemschutzmaske. Als neues Sicherheitsfeature ist auf der Rückseite eine rot leuchtende Positionsleuchte integriert, die auch in dunklen oder verrauchten Umgebungen erkennbar ist. Akzentuierende Fasen und Kanten sorgen für eine Emotionalisierung sowie eine charakteristische, markenbezogene Optik – und für ein Mehr an Stabilität.

A firefighting helmet is designed for extreme situations – its primary function is to protect the wearer. What's needed is therefore a combination of maximum sturdiness and optimal ergonomics. Weighing just 1.23 kilograms, the Heros H30 does more than just adapt to the size of the wearer's head: the position of the shell can also be adjusted so that the helmet doesn't slip and always remains balanced. In addition, functional add-ons are also available, including a helmet light that can be mounted without tools, a thermal imaging camera and a respirator mask. A red position light has been integrated into the back of the helmet as a new safety feature, ensuring visibility even in dark or smoke-filled surroundings. Accentuating bevels and edges emotionalise the design and give it the brand's characteristic look – as well as adding stability.

PETER STAUDINGER — PRODUKTMANAGER PERSÖNLICHE SCHUTZAUSRÜSTUNG, ROSENBAUER INTERNATIONAL AG

»Der Feuerwehrhelm funktioniert ein Stück weit als Ausdruck des persönlichen Status.«

»To some extent, the firefighting helmet serves as an expression of personal status.«

PETER STAUDINGER — **PRODUCT MANAGER PERSONAL PROTECTIVE EQUIPMENT, ROSENBAUER INTERNATIONAL AG**

→ **Der H30 ist nicht der erste Helm von Rosenbauer. Warum braucht es einen neuen Helm?**
Bei Rosenbauer befindet sich das Thema Feuerwehrhelm permanent in Entwicklung. Wir analysieren neue Materialien und Technologien, beobachten den Markt. So haben wir den Bedarf an einem besonders leichten Hochleistungshelm bereits erkannt, doch erst die verfügbaren Technologien gaben den Ausschlag für den Entwicklungsstart eines leichteren, kompakteren und funktionaleren Helms.

Wie wichtig ist es, dass ein Rosenbauer-Helm gleich als solcher erkannt wird?
Ein hoher Wiedererkennungswert ist eine markenstrategische Notwendigkeit. Für Rosenbauer als Familienunternehmen sind Produkte mit hoher gestalterischer Identifikationskraft von emotionaler Wichtigkeit. Die Wiedererkennbarkeit an sich ist mehrschichtig, Attribute wie der Kamm definieren ihn als Feuerwehrhelm. Die Formensprache mit dynamischen Linien und Kanten, die Betonung der Kopfform, die Logoplatzierung und Produktgrafik komplettieren die Rosenbauersche Design-DNA.

Welche Relevanz hat das Design bei der Entwicklung eines Sicherheitshelms?
Feuerwehrfrauen und -männer, egal ob hauptberuflich oder freiwillig aktiv, haben einen sehr großen emotionalen Bezug zur Feuerbekämpfung. Der Feuerwehrhelm funktioniert ein Stück weit als Ausdruck des persönlichen Status. Ergonomische Aspekte und Ergebnisse aus der Designentwicklung unterstützen die fehlerfreie Bedienung und somit die Sicherheit des Helms.

Welche Usability-Anforderungen standen im Vordergrund?
Der konsequente Leichtbau und die weit über die Normen hinausgehenden Sicherheitsmerkmale verbessern die Usability in jeder Situation. Bei der Usability im Sinne der aktiven Interaktion geht es um die intuitive, einfache und effiziente Bedienung, auch in Stress-Situationen. Usability bedeutet für uns aber auch die Anpassungsfähigkeit an den oder die Träger:in und an unterschiedliche Einsatzbereiche. Das zeigt die Vielzahl an kombinierbaren Zusatzoptionen unseres Online-Helm-Konfigurators.

Wann und wie wurden die künftigen Nutzer:innen involviert?
Rosenbauer pflegt intensiven und engen Kontakt zu seinen internationalen Kunden. Regelmäßige Umfragen und Studien dienen zur Identifizierung und Integration von Nutzerbedürfnissen. Nicht zuletzt legen wir Wert darauf, dass sich unsere Mitarbeiter:innen selbst bei der Feuerwehr engagieren. So entstehen Produkte für Feuerwehrfrauen und -männer von Feuerwehrfrauen und -männern.

Rosenbauer entwickelt und produziert Fahrzeuge, Löschtechnik, Ausrüstung und digitale Lösungen für Berufs-, Betriebs-, Werks- und freiwillige Feuerwehren. Mit rund 4.100 Mitarbeiter:innen und einer Präsenz in 120 Ländern ist Rosenbauer der weltweit größte Feuerwehrausstatter.

www.rosenbauer.com

→ **The H30 isn't Rosenbauer's first helmet. Why is a new helmet needed?**
At Rosenbauer, firefighting helmets are the subject of permanent development. We analyse new materials and technologies and monitor the market. As a result, we'd already identified the need for a particularly lightweight high-performance helmet, but it was only when the corresponding technologies became available that we could start developing a lighter, more compact and more functional product.

How important is it for a Rosenbauer helmet to be recognised as such immediately?
When it comes to brand strategy, strong recognition is a must. For Rosenbauer as a family business, products with a highly identifiable design are important at an emotional level. Recognisability per se is multilayered; attributes like the ridge define it as a firefighter's helmet, but it's things like the dynamic lines and edges of the design language, the accentuation of the head shape and the positioning of the logo and product graphics that constitute Rosenbauer's design DNA.

How relevant is the design to the development of a safety helmet?
Regardless of whether they're professionals or volunteers, firefighters have a very strong emotional bond with firefighting. To some extent, the firefighting helmet serves as an expression of personal status. Ergonomic aspects and the results of the design development help ensure correct use and improve the safety of the helmet as a result.

Which usability requirements were uppermost?
The consistent focus on lightness and the fact that the safety features go far beyond the norm improve usability in any situation. As for usability in the sense of active interaction, it's crucial to ensure intuitive, simple and efficient use, even in stressful situations. But for us, usability also means the helmet's capacity to adapt to its wearer and to different types of mission. That's evident from the multitude of combinable add-ons featured in our online helmet configurator.

When and how were future users involved?
Rosenbauer cultivates strong and close relationships with its international customers. Regular surveys and studies enable us to identify and integrate users' needs. And last but not least, we encourage our employees to be active in the voluntary fire brigade as well. That results in products for firefighters by firefighters.

Rosenbauer develops and produces vehicles, fire extinguishing systems, equipment and digital solutions for professional, industrial, plant and voluntary fire services. With approx. 4,100 employees and a presence in 120 countries, Rosenbauer is the world's largest provider of firefighting equipment.

www.rosenbauer.com

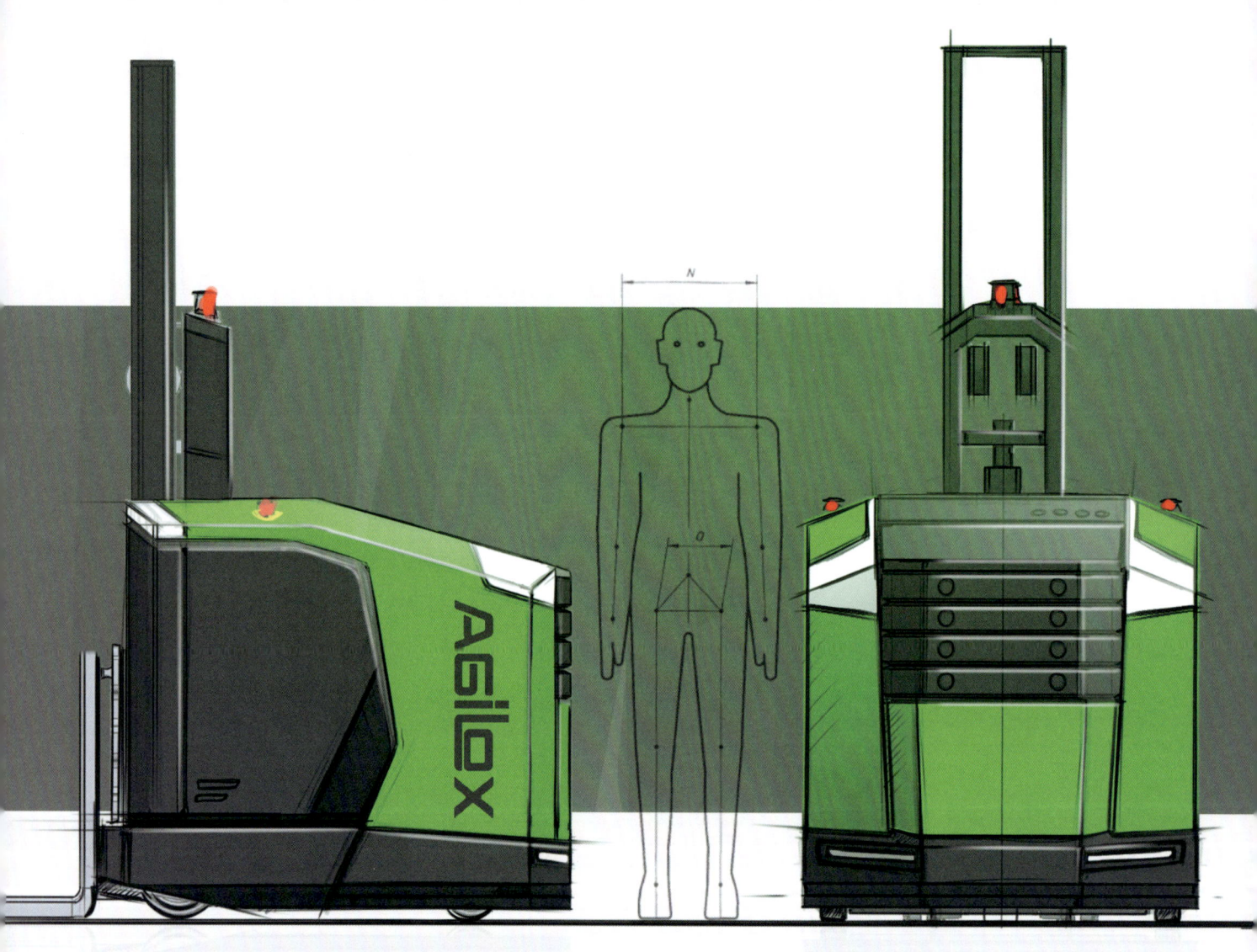

GOLD OCF **AUTONOMER GEGENGEWICHTSSTAPLER** 24
AUTONOMOUS COUNTERBALANCE FORKLIFT 25

JURY STATEMENT

Die Wirkung des Fahrzeugs ist smart und kraftvoll zugleich. Das Setzen von Leuchtelementen bringt ein emotionales Moment, das bewusst Analogien zu einem Gesicht erkennen lässt. Damit wird die Akzeptanz des Roboters erhöht. Trotz der technischen Dominanz zeigt sich der Stapler mit ausgewogenen Proportionen und einem unverwechselbaren Design. Ein Musterbeispiel für künftige, autonom agierende Roboter.

The vehicle makes a smart and powerful impact. The positioning of the illuminated elements adds an emotional aspect and exhibits deliberate similarities with a face, thereby increasing acceptance of the mobile robot. Although dominated by technical considerations, the forklift has nevertheless been designed with balanced proportions and a distinctive aesthetic. It sets an outstanding example for future autonomous robots.

HERSTELLER/MANUFACTURER
Agilox Services GmbH
Neukirchen bei Lambach
Österreich/Austria

DESIGN
Formquadrat GmbH
Linz
Österreich/Austria

VERTRIEB/DISTRIBUTOR
Agilox Services GmbH
Neukirchen bei Lambach
Österreich/Austria

Für die Intra-Logistik konzipiert, transportiert der 3,6 Tonnen schwere Stapler Paletten und andere Behältnisse autonom im Unternehmen. Das Fahrzeug hebt Lasten von 1,5 Tonnen bis zu 160 Zentimeter hoch und benötigt weder einen zentralen Leitrechner noch traditionelle Navigationshilfen, erschließt sich die Gegebenheiten vor Ort selbst, erkennt Hindernisse per Sensoren und umfährt sie. Konzipiert für den Mischbetrieb mit konventionellen Staplern, kommuniziert der OCF mit seinesgleichen und optimiert seine Route selbstständig. Mit einem omnidirektionalen Antrieb ausgerüstet, vermag der Stapler auch diagonal zu verfahren oder auf der Stelle zu drehen.

Das Design nimmt die markenspezifischen Vorgaben des Herstellers auf und dynamisiert das vergleichsweise große Fahrzeug durch ein geschicktes Setzen von Kanten, Fasen und Farben. Lichtelemente, über die mit der Umgebung kommuniziert wird, sind flächenbündig integriert.

Designed for intralogistics, the 3.6-tonne forklift autonomously transports pallets and other loads within the facility. The vehicle can lift loads of 1.5 tonnes to a height of up to 160 centimetres; it needs neither centralised control nor traditional navigation aids, learns about its surroundings on the job and uses sensors to recognise and avoid obstacles. Intended for mixed operation with conventional forklifts, the OCF communicates with its peers and optimises its route independently. Equipped with an omnidirectional drive, the lift is also able to move diagonally or turn on the spot.

The design incorporates the manufacturer's brand identity and gives the relatively large vehicle a dynamic look through the clever use of edges, bevels and colours. The illuminated elements that the forklift uses to communicate with its surroundings are fitted flush with the surface.

MARIO ZEPPETZAUER   **GESELLSCHAFTER, FORMQUADRAT GMBH**

»Wichtig ist, dass der Stapler Sicherheit und Verlässlichkeit vermittelt.«

»It's important for the forklift to convey safety and reliability.«

MARIO ZEPPETZAUER — PARTNER, FORMQUADRAT GMBH

→ **Der OCF ist autonom unterwegs – muss das Design dafür eine neue Sprache finden?**
Der OCF ist der zweite AMR (Autonomous Mobile Robot), den wir für Agilox gestalten durften. Dabei haben wir die Formensprache weiterentwickelt. Ein AMR soll seine Funktion nach außen erkennbar machen, dazu leistet schon die Hubgabel einen großen Beitrag. Rundherum bestehen viele Freiheiten durch den Wegfall der Führerkabine. Wichtig ist aber, dass das Produkt Sicherheit und Verlässlichkeit vermittelt – ein AMR darf nicht aussehen wie aus einem Science-Fiction-Film. Durch die dezente Formgestaltung mit einem Gesicht, das von den Signalleuchten flankiert wird, erweckt der OCF einen freundlichen, aber selbstbewussten Eindruck.

Gehen Sie an einen Roboter gestalterisch anders heran als an ein konventionelles Produkt?
Der Kontext eines Produkts muss bei der Gestaltung immer mitgedacht werden, auch bei einem autonomen Roboter. Dazu gehören zum Beispiel das Produktumfeld, die User Experience, die Materialien, die Fertigungsmethoden. Eigenständig agierenden Produkten schreibt der Mensch dennoch gern eine Persönlichkeit zu, daher darf der OCF weder aggressiv noch unbeholfen wirken. Das muss im Designprozess mitgedacht werden. Am Ende muss die Gestaltung zur Marke passen, zu vernünftigen Preisen produzierbar sein und natürlich auch Käufer:innen ansprechen.

Wie kommuniziert der OCF mit seinen menschlichen Kolleg:innen?
Die Kommunikation zwischen Mensch und Roboter findet auf verschiedenen Ebenen statt. Bedient wird via App, einige wenige Funktionen befinden sich auch als Bedienknöpfe am Roboter selbst. Im Betrieb ist es wichtig, dass der Roboter vor allem wahrnehmbar ist. Das erreichen wir durch großflächige Signalleuchten, die aus allen Richtungen gut sichtbar sind, den Betriebszustand anzeigen und vorausgehende Aktionen kommunizieren. Der Floor Spot illuminiert den vor dem Stapler liegenden Weg und zeigt, wohin die Fahrt geht. Auch akustisch macht der Stapler auf sich aufmerksam. Die logistischen Arbeiten erledigt der OCF selbstständig, im Normalbetrieb ist keine Kommunikation notwendig.

Investitionsgüter sind langlebig – was bedeutet dies für Ihre Konzeptionsansätze?
Die Produkte von Agilox sind hochkomplex. Die technische Vielschichtigkeit muss für Nutzer:innen in Bedienung und Form vereinfacht werden, die Wartung muss mühelos möglich sein und Einzelkomponenten müssen im Falle eines Schadens rasch austauschbar sein. Diese Prämissen vereint der OCF durch eine schlichte und reduzierte Gestaltung, die auch noch in 15 Jahren ihre Strahlkraft behält. Mit Farbkontrasten sowie wenigen, aber bewusst gestalteten Lichtkanten wird Spannung geschaffen, die den OCF wiedererkennbar und aufregend macht.

Gegründet von Stefan Degn und Mario Zeppetzauer, steht Formquadrat mit Standorten in Linz und Gmunden seit über 20 Jahren für die Gestaltung technischer Produkte, die vielfach ausgezeichnet und in den Märkten erfolgreich sind.

www.formquadrat.com

→ **The OCF moves around autonomously – does the design have to find a new language for that?**
The OCF is the second AMR (Autonomous Mobile Robot) that we've designed for Agilox, and we've continued to evolve the design language. An AMR should make its function recognisable to the outside world, and the lifting fork goes a long way towards doing that. And the fact that there's no need for a driver's cab results in a lot of freedoms in other respects. However, it's important for the product to convey safety and reliability – an AMR definitely shouldn't look like something from a science fiction film. Thanks to the understated design complete with a face that's flanked by the signal lights, the OCF makes a friendly but self-confident impression.

Do you take a different approach to designing a robot than you would to designing a conventional product?
When you're designing a product you always have to keep its context in mind, and that goes for an autonomous robot too. That includes things like the product environment, the user experience, the materials and the production methods. Nevertheless, people tend to attribute autonomous products with a personality, which is why it's vital that the impression the OCF makes is neither aggressive nor clumsy. That has to be factored into the design process. At the end of the day the design has to be a fit with the brand, producible at a reasonable price and appeal to its buyers.

How does the OCF communicate with its human colleagues?
The communication between human and robot takes place at various levels. It's controlled via an app, and there are a few buttons on the robot itself for certain functions. When it's in operation, the most important thing is for the robot to be noticeable. We achieve that via large signal lights that are visible from any direction, indicate the operating status and communicate imminent actions. The floor spot illuminates the path in front of the forklift and shows what direction it's headed in. And it uses acoustic signals to draw attention to itself as well. The OCF performs logistics-related tasks autonomously, so no communication is necessary when it's in standard operating mode.

Capital goods have a long service life – what does that mean for your design approach?
The products Agilox makes are highly complex. The control and design of their technical sophistication has to be simplified so as to be accessible to users, maintenance has to be effortless and it has to be possible to replace damaged components quickly. The OCF resolves those issues with a simple and reductive design that will be just as charismatic 15 years from now. The colour contrasts and the few but deliberately positioned contours create a sense of suspense that makes the OCF both recognisable and intriguing.

Founded by Stefan Degn and Mario Zeppetzauer and with locations in Linz and Gmunden, Formquadrat has been designing award-winning and commercially successful technical products for more than 20 years.

www.formquadrat.com

INVESTITIONSGÜTER, WERKZEUGE
CAPITAL GOODS, TOOLS

28
29

FOCUS GOLD

MOBILER BATTERIE— SPEICHER

GOLD

INSTAGRID ONEMAX
MOBILER BATTERIESPEICHER / PORTABLE STORAGE BATTERY

JURY STATEMENT

Trotz seines Gewichts von 20 Kilogramm wirkt der Stromspeicher optisch leicht und schlank, wobei es trotzdem gelingt, sein enormes Kraftpotenzial zum Ausdruck zu bringen. Die Bedienfront ist klar strukturiert und damit intuitiv nutzbar, auch in rauen Umgebungen. Nicht zuletzt lässt sich die Produktgrafik als ausgesprochen professionell bezeichnen.

Despite weighing 20 kilograms, the storage battery manages to look light and slender while nevertheless expressing its huge potential for supplying power. The user interface is clearly structured and therefore intuitive, even in harsh environments. Last but not least, the product graphics are extremely professional.

HERSTELLER/MANUFACTURER
Instagrid GmbH
Ludwigsburg

DESIGN
Inhouse
Felix Fuchs

VERTRIEB/DISTRIBUTOR
Instagrid GmbH
Ludwigsburg

Wer eine mobile Stromversorgung benötigt, kam bislang kaum an einem benzingespeisten Generator vorbei. Das ändert sich nun mit dem Batteriespeicher, der 2,1 kWh Energie als Wechselstrom bei 230 Volt Spannung ausgibt. Mit einer Spitzenleistung von 18.000 Watt kann der Speicher sogar Schweißgeräte oder Plasmaschneider speisen.

Beim Betrieb entstehen keine Emissionen, der Speicher arbeitet wartungsfrei, die Lithium-Ionen-Batterien lassen sich in kurzer Zeit wieder aufladen. Der Ladestatus lässt sich vom runden, in mehrere Segmente aufgeteilten LED-Interface auch aus größerer Entfernung klar ablesen. Von einem Rohrkäfig zusätzlich geschützt, ist das Gehäuse aus Recycling-Alu auf größte Robustheit ausgelegt. Bei der Konzeption wurde zudem viel Wert auf einfache Reparaturfähigkeit gelegt – einzelne Module lassen sich schnell austauschen. Als Schnittstelle zum Verbrauchergerät dient eine konventionelle Steckdose.

Up until now, a mobile power supply almost inevitably meant a petrol generator. That is about to change thanks to this storage battery, which provides 2.1 kilowatt hours of energy in the form of alternating current with an output voltage of 230 volts. With peak power of 18,000 watts, the battery can even power welding tools or plasma cutters.

The system, which is based on lithium-ion batteries, produces no emissions, is maintenance-free and can be recharged within a short space of time. Even from a considerable distance, the charge status is easy to read thanks to the round LED interface, which is subdivided into multiple segments. Enclosed in a metal cage for extra protection, the housing is made of recycled aluminium and designed for maximum robustness. The concept also attaches great importance to repairability – the individual modules are easy to replace. A conventional socket serves as the interface to the electrical load.

FELIX FUCHS — DESIGN UND UX, INSTAGRID GMBH

»Gleich zu Anfang stehen schnelle Mockups und Prototypen, die direkt zum Kunden gehen. Testen. Scheitern. Neu denken.«

»Right at the outset we make quick mockups and prototypes that go straight to the customer. Test. Fail. Think again.«

FELIX FUCHS

DESIGN AND UX, INSTAGRID GMBH

→ **Ist OneMax das erste Produkt von Instagrid? Wird es weitere Produkte geben?**
OneMax ist das erste Produkt, das unter der Marke Instagrid am Markt ist. Nach dem großartigen Feedback fokussieren wir uns aktuell darauf, Steckdosen und Spannungen auf die unterschiedlichen Anforderungen weltweit anzupassen. Vor kurzem haben wir zudem eine Variante speziell für Behörden und Organisationen wie Feuerwehr oder THW vorgestellt. Außerdem stehen Tools zur Verwaltung von Geräteflotten auf der Entwicklungsagenda.

Wie relevant ist das Design für Ihr Standing am Markt?
Um uns am Markt zu positionieren, spielt das Design eine essenzielle Rolle. Wir haben von Anfang an die Anforderungen unserer Kund:innen in den Designprozess eingebunden. Das Ergebnis ist ein Produkt, das nicht nur durch technologische Überlegenheit überzeugt, sondern sich auch durch sein eigenständiges Design von Mitbewerbern abhebt. Wobei gutes Design nicht nur die äußere Erscheinung meint – Funktionalität, Bedienbarkeit und Nachhaltigkeit sind ebenfalls wichtige Themen.

In welchem Stadium der Entwicklung kam das Design hinzu?
Instagrid verfolgt einen designgetriebenen Ansatz. Gleich zu Anfang stehen schnelle Mockups und Prototypen, die direkt zum Kunden gehen. Testen. Scheitern. Neu denken. Insofern ist Design bei uns ein Grundstein für jede Entwicklung und geht Hand in Hand mit der technologischen Entwicklung.

Wo lagen die Herausforderungen beim Design und der Konzeption?
Davon gab es einige. Angefangen vom Fehlen einer »etablierten« Designsprache für mobile Batteriespeicher bis zur Herausforderung, Kund:innen, die bisher nur klassische, benzingespeiste Generatoren kannten, mit einem hinsichtlich Design, Leistung und Handhabung innovativen Produkt zu überzeugen. Letztendlich auch die Aufgabe, unser Produkt so umweltverträglich wie möglich zu gestalten.

Welche Bedeutung kommt der Gestaltung des Interfaces zu – in Sachen Bedienbarkeit, aber auch Branding?
Mit der Gestaltung unseres Interfaces haben wir uns sehr bewusst gegen viele Mitbewerber positioniert, die mit etlichen Steckmöglichkeiten und anfälligen Displays die Nutzer:innen überfordern. Wir setzen auf klare Ablesbarkeit, einen auf die Essenz reduzierten Drehschalter als Bedienelement sowie auf eine konventionelle und damit universelle Steckdose.

Wie wichtig sind Designauszeichnungen für Sie?
Als junges Unternehmen, dessen Fokus stark im B2B-Sektor liegt, ist es sehr wichtig für uns, wahrgenommen zu werden und die Marke zu etablieren. Designauszeichnungen sind da natürlich ein ausgezeichnetes Mittel. Ebenso wichtig für uns ist das Feedback einer qualifizierten Jury – wie werden Features und Designelemente wahrgenommen? Wir sehen diesen Input als Chance für Verbesserungen und für die Weiterentwicklung unserer Designsprache.

> Instagrid wurde 2018 in Ludwigsburg von Andreas Sedlmayr und Sebastian Berning gegründet. Das Unternehmen hat sich auf innovative Lösungen für Batteriespeichertechnologie spezialisiert und ist europäischer Marktführer für tragbare Hochleistungsbatteriesysteme. Momentan arbeiten in Standorten in Deutschland, Finnland und Großbritannien rund 85 Mitarbeitende aus über 20 Ländern.
>
> www.instagrid.co/de

→ **Is OneMax Instagrid's first product? Will there be more?**
OneMax is the first product to be marketed under the Instagrid brand. After the fantastic feedback, we're now focusing on adapting the sockets and voltages to the various requirements in different parts of the world. We also recently presented a variant specially for authorities and organisations like the fire brigade or the Federal Agency for Technical Relief (THW). And tools for managing device fleets are on the development agenda too.

How relevant is design for your market standing?
Design plays an essential role in establishing our positioning. Right from the start, we've always incorporated our customers' requirements into the design process. The result is a product that isn't just compelling for its technological superiority but that stands out from competitors due to its original design as well. Although good design certainly doesn't just mean what something looks like on the outside – functionality, usability and sustainability are just as important.

At what stage of development did design come into the picture?
Instagrid takes a design-driven approach. Right at the outset we make quick mockups and prototypes that go straight to the customer. Test. Fail. Think again. So as far as we're concerned, design is a cornerstone for every development and goes hand in hand with the technological side of development.

Where did the challenges lie in terms of the design and concept?
There were several – starting with the absence of an »established« design language for portable storage batteries, all the way to the challenge of convincing customers who have only known classic, petrol-fuelled generators up until now of an innovative product whose design, performance and handling break new ground. And last but not least, the mission to make our product as environment-friendly as possible.

How important is the design of the interface – not just in terms of usability but with regard to branding too?
The design of the interface is a very deliberate attempt to position ourselves in stark contrast to a lot of our competitors, who overwhelm users with too many socket options and make their displays too fragile. Our focus was on good legibility, using a dial as the control element and reducing it to the essentials, and equipping the battery with a conventional and therefore universal socket.

What role do design awards play for you?
As a young company with a strong focus on the B2B sector, it's very important for us to be noticed and to establish the brand. And design awards are obviously an excellent tool in that respect. But the feedback from a qualified jury is just as important to us – how are the features and design elements perceived? We see that input as an opportunity to make improvements and evolve our design language.

> Instagrid was founded in Ludwigsburg in 2018 by Andreas Sedlmayr and Sebastian Berning. The company specialises in innovative solutions for battery technology and is the European market leader for portable high-performance battery systems. Around 85 employees from 20 countries currently work at its locations in Germany, Finland and the United Kingdom.
>
> www.instagrid.co/de

SILVER FTS **AUTONOMER GEGENGEWICHTSSTAPLER**
→ SEITE/PAGE **AUTONOMOUS COUNTERBALANCE FORKLIFT**
42

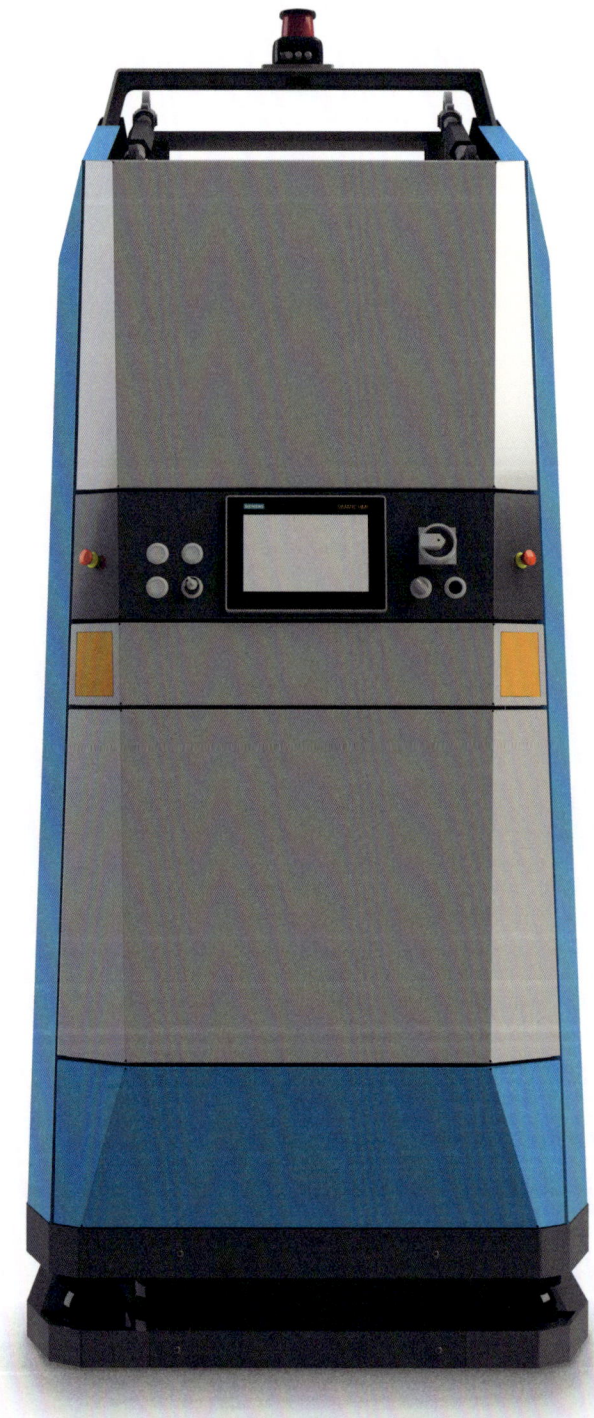

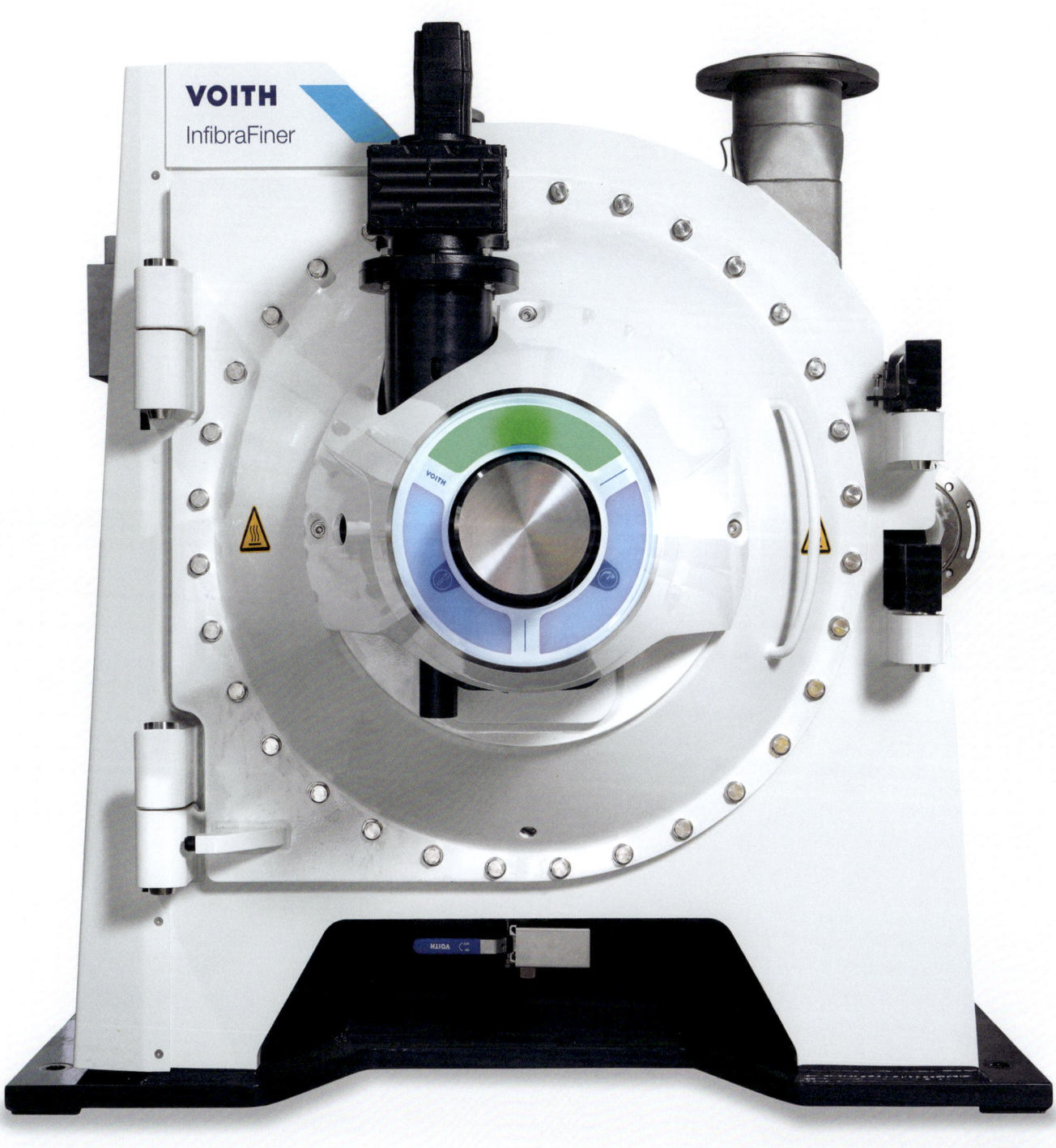

SILVER	MK	KONTAKTGEBER
	→ SEITE/PAGE	CONTACT BLOCKS
	44	

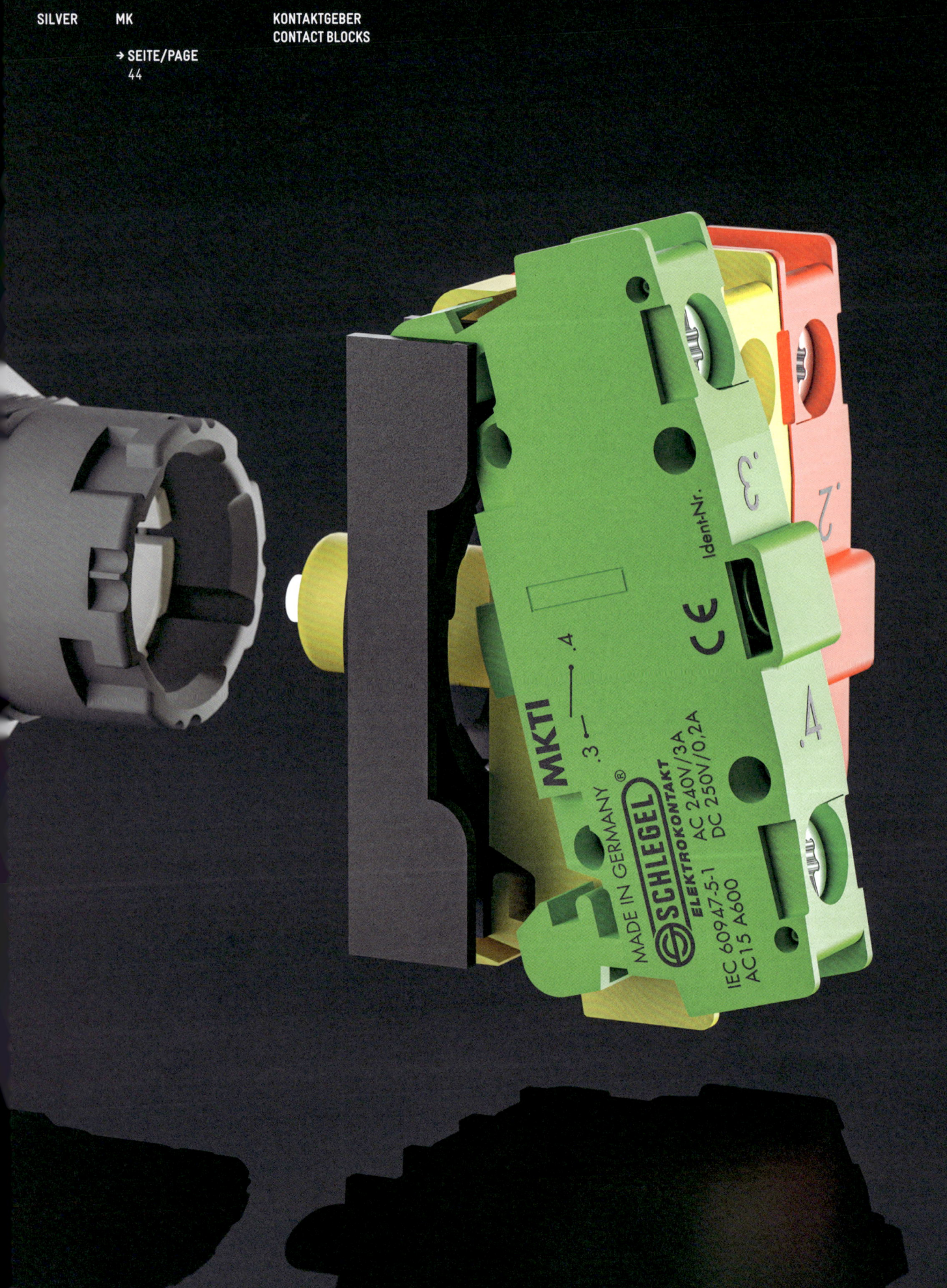

SILVER

DUOXPAND

LANGSCHAFTDÜBEL
FRAME FIXING

JURY STATEMENT

Der zweifarbige Dübel zeigt eine sehr progressive Produktgrafik – sie ist einerseits funktional begründet, aber verweist klar auf das Corporate Design des Unternehmens. Damit differenziert sich der Dübel am Markt, visualisiert sein besonderes Eigenschaftsprofil und seinen professionellen Wert.

The product graphics of the two-coloured plug are very progressive – as well as being function-based, they are also a clear reflection of the company's corporate design. As a result, the fixing stands out on the market and visualises both its special characteristics and professional merits.

HERSTELLER/MANUFACTURER
Fischerwerke GmbH & Co. KG
Waldachtal

DESIGN
Teams Design GmbH
Esslingen

VERTRIEB/DISTRIBUTOR
Fischer Deutschland Vertriebs GmbH
Waldachtal

Rationelles Arbeiten ist für Profis ein Muss – besonders im Bausektor, auch beim Setzen von Befestigungspunkten im Mauerwerk mit Hilfe von Dübeln. Nun ist Mauerwerk nicht gleich Mauerwerk, sondern kann sehr unterschiedlich aufgebaut sein – darauf sollte der Dübel abgestimmt sein. Damit der »Dübelzoo« im Handwerkerlager nicht allzu sehr ausufert, entwickelt der Hersteller Dübel, die sich für mehrere Untergründe zugleich eignen. So lässt sich der neue DuoXpand mit seiner speziellen Lamellengeometrie sowohl in Massiv- als auch Lochsteinmauerwerk einsetzen, wobei letzteres den Dübel veranlasst, sich hinter dem Steinsteg aufzuspreizen und mit Hinterschnitten zu verankern. Das an seiner roten Farbe erkennbare Spreizelement ist für zwei Verankerungstiefen ausgelegt, was den Dübel nochmals universeller macht. Die bis 230 Millimeter Länge erhältlichen Dübel werden zeitsparend mit vormontierter Schraube geliefert.

Systematic working is a must for professionals – especially in the construction sector, and including when it comes to anchoring fixings in concrete and masonry frames with the aid of plugs. However, the makeup of such frames can vary considerably – and the plug needs to be selected accordingly. To avoid the number of different plugs that construction firms have to keep in stock spiralling out of control, the manufacturer develops products that are equally suitable for several types of substrate. Thanks to its special lamella geometry, for instance, the new DuoXpand can be used for both solid and hollow masonry; in the latter case, the plug expands behind the web and anchors itself with undercuts. Recognisable by its red colour, the expanding element is designed for two anchorage depths, making the plug even more universal. Available in lengths up to 230 millimetres, the plugs are supplied with a pre-mounted screw for time-saving installation.

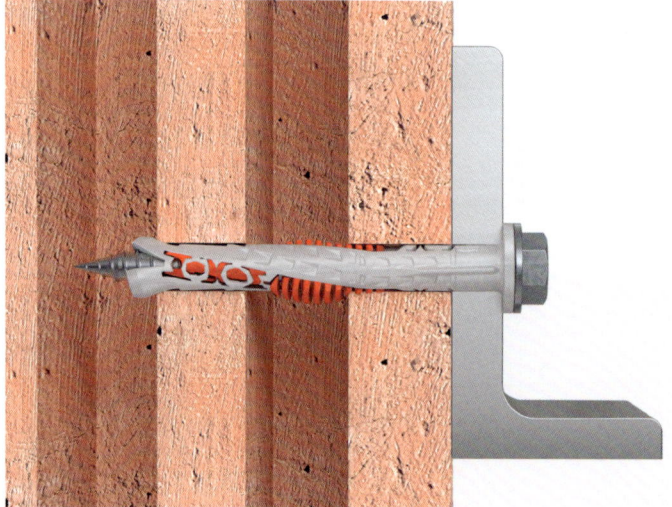

SILVER — STE300 — DECKEN- UND MONTAGESTÜTZE / CEILING AND ASSEMBLY SUPPORT

JURY STATEMENT

Ein solide gestaltetes Tool für den handwerklichen Alltag. Die Bedienung ist schnell und selbsterklärend, die Elemente kompakt und robust im Gebrauch, die ergonomische Ausformung des Griffs durchdacht. Eine ideale Hilfe für jede Art von Montage.

A solidly designed tool for everyday construction work. The handling is quick and self-explanatory, the individual elements are compact and hardwearing, the ergonomic design of the handle has been well thought through. An ideal aid for all sorts of assembly.

HERSTELLER/MANUFACTURER
Bessey Tool GmbH & Co. KG
Bietigheim-Bissingen

DESIGN
Weinberg & Ruf
Filderstadt

VERTRIEB/DISTRIBUTOR
Bessey Tool GmbH & Co. KG
Bietigheim-Bissingen

Montagestützen mit variabler Auszugslänge sind bei der Überkopfmontage von Verkleidungen oder Tragwerken unverzichtbare Hilfen. Wichtig dabei ist die möglichst einfache, schnelle und sichere Bedienung. Unter diesen Prämissen wurden die STE-Stützen in drei Größen entwickelt. Allen gemeinsam ist das mehrstufige Verstellprinzip: Mittels einer Schnellverschiebetaste wird die Stahlrohrkonstruktion grob vorteleskopiert, dann per Pumphebel weiter ausgefahren, bis die Anlageplatte anliegt. Schließlich wird durch das Drehen des Griffs die gewünschte Haltekraft eingestellt. All dies erfolgt praxisgerecht einhändig am ergonomisch geformten Griff.

Das Modell STE300 lässt sich bis drei Meter ausziehen, die Tragfähigkeit beträgt dann 110 Kilogramm. Die schwenkbaren Boden- und Anlageplatten mit ihren rutschsicheren Oberflächen ermöglichen Stützwinkel von bis zu 45 Grad.

Telescopic assembly supports are indispensable when mounting panels or structural frames overhead and need to be as simple, fast and safe to use as possible. These requirements were the basis for the development of the STE supports. Available in three sizes, they all share the same multi-stage adjustment principle: using a quick-slide button, the tubular steel support is extended to the approximate length required and then adjusted by means of a pump lever until the support plate makes contact. Finally, the required holding strength is set by turning the grip. Thanks to the ergonomically shaped handle, all these steps can be conveniently executed using one hand.

When fully extended to a length of three metres, the STE300 model can support a load of 110 kilograms. The swivelling floor and support plates are equipped with non-slip surfaces and allow the props to be set at angles of up to 45 degrees.

SILVER FTS AUTONOMER GEGENGEWICHTSSTAPLER
AUTONOMOUS COUNTERBALANCE FORKLIFT

JURY STATEMENT

Ein Paradebeispiel für die Leistung von Designverantwortlichen, betrachtet man die rein technisch entwickelte Variante davor. Das Unternehmen hat hier einen großen Schritt gemacht und zeigt den Willen zu einem eigenständigen Markendesign. Die Formensprache entwickelt sich aus den verfügbaren Fertigungsverfahren und ist durchaus zeitlos.

A prime example of what designers can contribute – especially as compared to the predecessor model, which was developed from a purely technical perspective. The company has taken a big step in this respect and is demonstrating the will to create a distinct design for its brand. The design language has been derived from the production processes available and is definitely timeless.

HERSTELLER/MANUFACTURER
Bleichert Automation GmbH & Co. KG
Osterburken

DESIGN
Ottenwälder und Ottenwälder
Schwäbisch Gmünd

VERTRIEB/DISTRIBUTOR
Bleichert Automation GmbH & Co. KG
Osterburken

Gedacht für die autonome Intralogistik, vermag der kompakte Stapler dank seines Gewichts von fünf Tonnen auch schwere Lasten zu transportieren und zu heben. Mittels des Laserscanners auf der Sensorbrücke erstellt der Stapler eine Karte seiner Arbeitsumgebung und navigiert selbstständig durch die Produktions- oder Lagerbereiche. Weitere, im Chassis formal integrierte Scanner erkennen auch Hindernisse im Bodenbereich. Da eine hohe Eigenfertigungsquote gefordert war, nutzt die Gestaltung die Blechbearbeitungsfähigkeiten des Herstellers aus und integriert notwendige Fugen geschickt in das facettierte Gesamtbild; die Neigung der Front visualisiert klar die Fahrtrichtung. Die Hülle verzichtet weitgehend auf exponierte Elemente, integrierte Blinkleuchten an Vorder- und Rückseite erhöhen die Sicherheit. Für den formal geschickten Übergang zum Hubmast hin sorgen prismatische Blenden.

Intended for autonomous intralogistics, the forklift weighs 5 tonnes and is thus able to transport and lift heavy loads despite its compact design. Using the laser scanner on the sensor bridge, the forklift creates a map of its work environment and navigates the production or warehouse halls autonomously. Other scanners integrated into the chassis even detect obstacles on the floor. Because a high degree of in-house production was called for, the design utilises the manufacturer's sheet metal processing capabilities and cleverly integrates the necessary joins into the faceted overall shape; the sloping front clearly visualises the direction of travel. The housing largely avoids protruding elements, integrated flashing lights on the front and back increase safety. Prismatic mouldings create a clever transition to the lift mast.

SILVER   **INFIBRAFINER**   **PAPIERFASER-AUFBEREITER / PAPER FIBRE REFINER**

> **JURY STATEMENT**
>
> Hier wurde eine Maschine direkt gestaltet, also ohne additive Verkleidungen zu nutzen. Dabei zeigt das Design durchaus die massive Konstruktionsweise, die wiederum für hohe Zuverlässigkeit steht. Interessant ist die runde und segmentierte Statusanzeige, die um die zentrale Achse der Maschine platziert wurde.
>
> The machine has been conceived in a very direct way without the use of additive panelling. As a result, the design reveals its imposing construction, which in turn signifies a high degree of reliability. The round, segmented status display positioned around the central axle of the machine is a particularly interesting feature.

HERSTELLER/MANUFACTURER
Voith Paper
Ravensburg

DESIGN
Defortec GmbH
Dettenhausen

VERTRIEB/DISTRIBUTOR
Voith Paper
Ravensburg

Die massive Maschine spielt im Kontext des Papierrecyclings eine wichtige Rolle: Im Inneren wird das Rohpapier gemahlen und seine Faserstruktur für die Weiterverarbeitung angeraut. Da die Maschine in einer schmutzintensiven Umgebung arbeitet, sind alle Elektronikkomponenten und Verkabelungen in das Gehäuse integriert. Glatte Flächen und weiche Formen vereinfachen die Reinigung des Geräts.

Der charakteristisch konische Frontbereich schützt die komplexe Steuerungssensorik und dient auch als Aufnahme für das sogenannte SmartLight. Diese neu entwickelte Statusanzeige kommuniziert über ihre farbigen Kreissegmente die Betriebszustände oder Wartungsanforderungen – auch über größere Entfernungen. Außerdem erhält die Anlage eine herstellerspezifische optische Signatur.

The massive machine plays an important role in the context of paper recycling: inside it, the raw paper is ground and its fibre structure roughened for further processing. Because the machine works in a dirt-intense environment, all the cabling and electronic components are integrated into the housing. Smooth surfaces and soft forms simplify cleaning.

The distinctive conical front section protects the complex control sensors and also incorporates the SmartLight – a newly developed status indicator that uses coloured segments to communicate operating modes or maintenance requirements, even over considerable distances. In addition, the machine conveys the visual identity of its manufacturer.

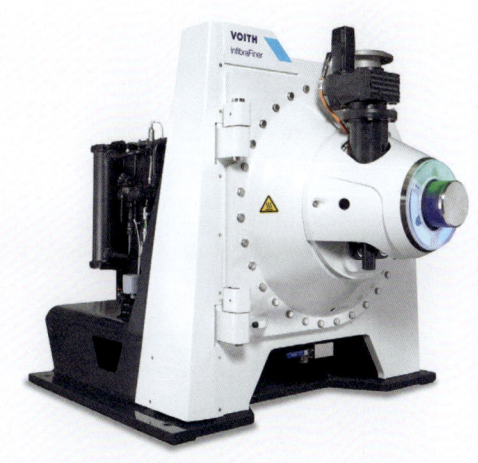

SILVER MK KONTAKTGEBER / CONTACT BLOCKS

> **JURY STATEMENT**
>
> Obwohl die Kontaktgeber quasi unsichtbar agieren, sind sie sehr gut gestaltet. Farbgebung und Anschlussposition erleichtern die Installation erheblich. Insgesamt erlaubt diese Bauweise die Realisierung schlankerer Panels. Ein gekonntes Beispiel für Komponentendesign.
>
> Even though the contacts operate out of sight, they are very well designed. The colour coding and position of the screw terminal make installation considerably easier. All in all, the design allows for shallower control panels. A masterly example of component design.

HERSTELLER/MANUFACTURER
Georg Schlegel GmbH & Co. KG
Dürmentingen

DESIGN
Inhouse

VERTRIEB/DISTRIBUTOR
Georg Schlegel GmbH & Co. KG
Dürmentingen

Taster oder andere manuell-analoge Bedienelemente von Maschinen wirken auf rückseitig platzierte Kontaktgeber, die für die eigentliche elektrische Schaltfunktion sorgen. Dafür braucht es Platz unter der Oberfläche eines Bedienpanels, der allerdings nicht immer ausreichend vorhanden ist. Für diese Fälle steht nun eine Reihe kompakter Kontaktgeber bereit, deren Einbautiefe auf unter 18 Millimeter begrenzt ist, aber die üblichen Leistungsdaten bietet. Dank der kleineren Bauweise wird weniger Material benötigt. Die farbig differenzierten Schließer, Öffner und LED-Beleuchtungen werden bedarfsgerecht variabel im zugehörigen Modulhalter kombiniert. Die neuen Kontaktgeber sind kompatibel mit den bekannten Betätigersystemen des Herstellers, die Kabelfixierung erfolgt aktuell per Schraubanschluss.

Pushbuttons and other manual analogue control elements for machines actuate contacts positioned behind them; it is these contacts that are responsible for the actual function of an electrical switch. They have to be housed under the surface of the control panel, where the space available is sometimes in short supply. For cases such as these, the manufacturer has developed a series of compact contact blocks that need a recess depth of less than 18 millimetres while still providing all the usual performance characteristics. Thanks to their smaller size, less material is required. The colour-coded contact blocks for normally open, normally closed and and LED illumination can be combined in the corresponding module holder as required. The new contact blocks are compatible with the manufacturer's familiar actuator systems; the wiring is currently fixed in place by a screw terminal.

SPECIAL MENTION GENIUSPOWER STROMVERSORGUNGSEINHEIT
POWER SUPPLY UNIT

> **JURY STATEMENT**
>
> Dass Design immer auch die Details im Blick hat, zeigt sich gerade an diesem flugzeugtechnischen Produkt. Die alles dominierende Leichtbauforderung wurde durch die Reduktion und Vereinfachung der Teile geschickt gelöst. Notwendige Lüftungsöffnungen verwandeln sich in eigenständige Designelemente, die Markenidentität ist klar erkennbar.
>
> Design is always mindful of the details – as demonstrated by this technical product for the aircraft industry. The all-dominating imperative of lightweight construction has been cleverly resolved by reducing and simplifying the components. The openings necessary for ventilation have been transformed into original design elements, the brand identity is clearly recognisable.

HERSTELLER/MANUFACTURER
KID Systems GmbH
Buxtehude

DESIGN
Zweigrad GmbH & Co. KG
Hamburg

VERTRIEB/DISTRIBUTOR
KID Systems GmbH
Buxtehude

Im Flugzeugbau ist jedes Gramm erspartes Gewicht ein wertvolles Argument – dies gilt selbst für kleine Bauteile, die unsichtbar ihren Dienst tun. Dazu gehören beispielsweise auch jene Geräte, die, in den Passagiersitzen integriert, die Energieversorgung für mitgebrachte Geräte übernehmen. Das Design des neu konzipierten GeniusPower sollte einerseits die Markenwerte transportieren, andererseits waren die Spielräume technisch und ökonomisch sehr begrenzt.

Das Ergebnis ist ein neues Gehäuse aus blau eloxiertem Aluminium, das mit seinen großen Eckradien und dem individuellen Lochraster ganz eigenständige Zeichen setzt. Aus nur zwei Biegeblechteilen bestehend, reduziert das Designkonzept obendrein die Herstellungskosten gegenüber dem rein technisch entwickelten Vorläufermodell. Das Brand-Logo, notwendige Informationen und Warnhinweise werden per Lasergravur perfekt reproduzierbar umgesetzt.

In aircraft construction, every gram of weight saved is a valuable argument. That applies even to small components that do their job invisibly, such as the devices that are integrated into the cabin seats and supply power for the electronics passengers bring on board with them. On the one hand the design of the new GeniusPower was to communicate the maker's brand values, on the other the degree of technical and economic freedom was very limited.

The result is a new housing made of blue anodised aluminium that uses large-radius corners and a distinctive perforated pattern to create a very original look. What's more, because it consists of just two bent sheet parts, the design concept also reduces production costs as compared to the predecessor model, which was developed based on purely technical considerations. The brand logo, important information and warnings are laser-engraved and can therefore be reproduced in perfect quality.

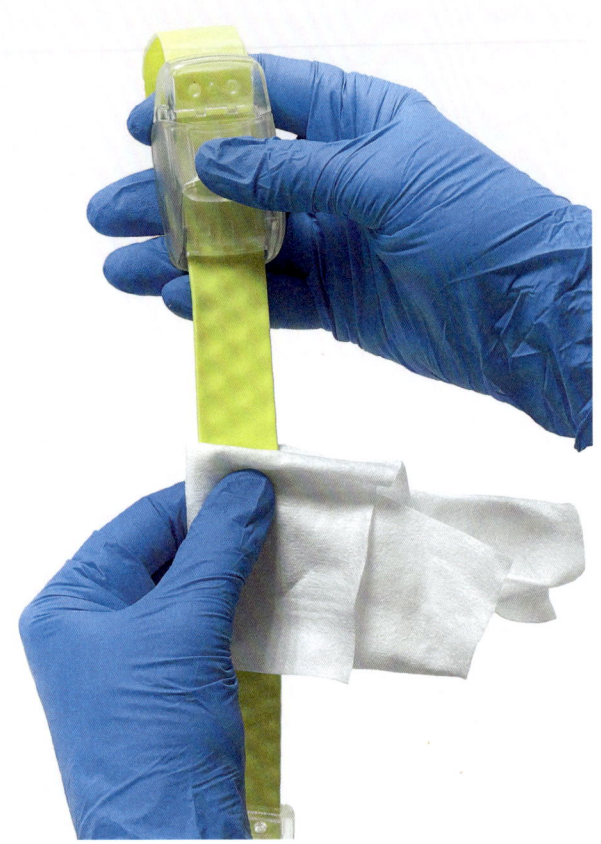

1 → SEITE/PAGE
50, 51

HEALTHCARE
HEALTHCARE

SPECIAL MENTION:
1 CBC NUCOS
Kimetec GmbH
Ditzingen

Wer medizinische Geräte gestaltet, bewegt sich in einem besonders sensiblen Bereich und übernimmt große soziale Verantwortung für Patient:innen wie auch für das medizinische Personal. Interessant dabei ist, dass nicht nur die Akutmedizin oder die Diagnostik von durchdachtem Design profitieren, sondern auch scheinbar selbstverständliche Hilfsmittel des medizinischen Alltags, die eine neue Qualität erhalten.

Designers of medical equipment operate in a particularly sensitive area and have a high level of social responsibility towards patients and medical staff alike. It's interesting to note that acute medicine or diagnostics aren't the only areas where well-conceived design can make a difference: it benefits even seemingly self-evident medical utensils for day-to-day use as well by imparting a whole new quality.

SPECIAL MENTION | CBC NUCOS → SEITE/PAGE 51 | VENENSTAUER TOURNIQUET

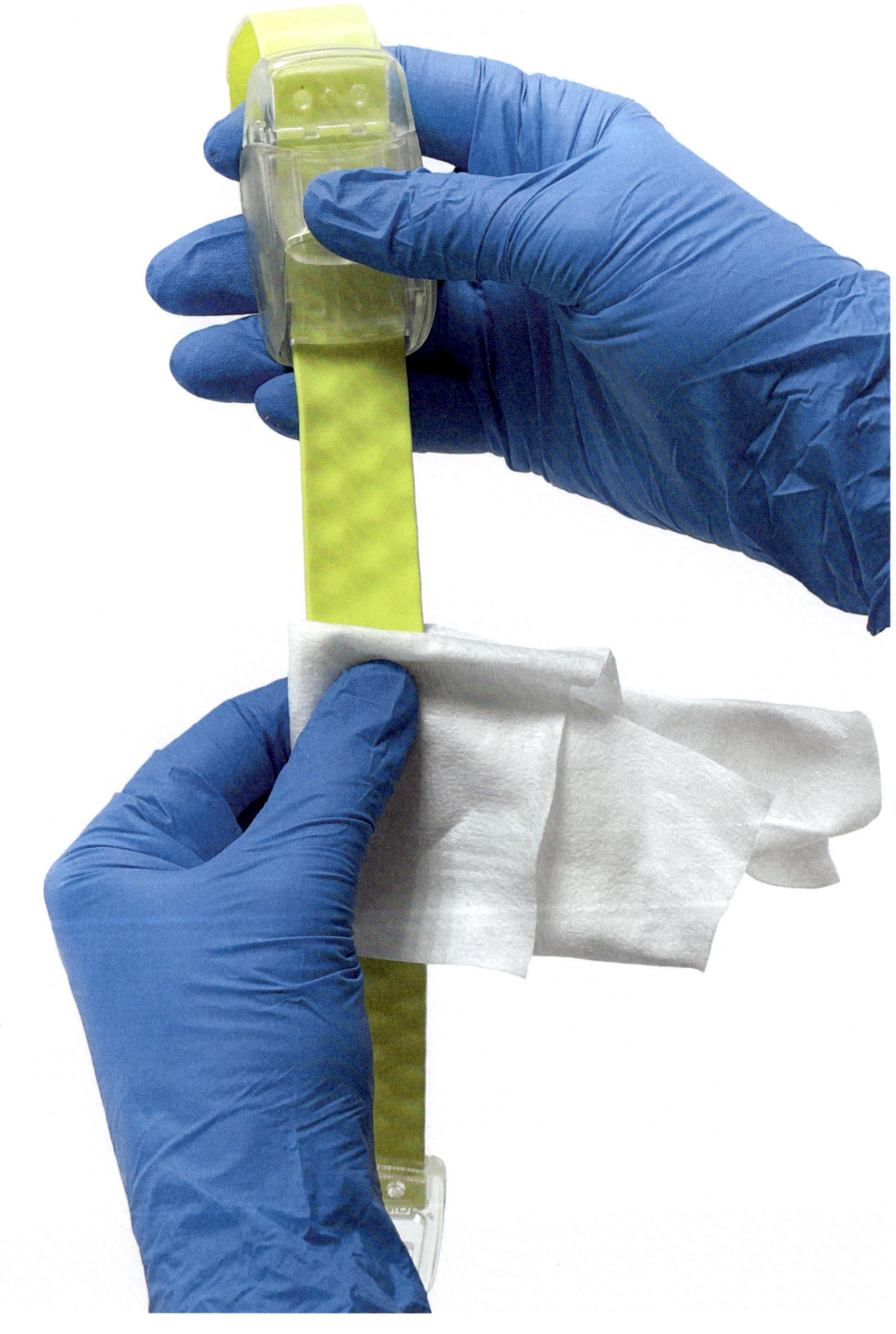

SPECIAL MENTION — CBC NUCOS — VENENSTAUER / TOURNIQUET

JURY STATEMENT

Warum eigentlich kam noch niemand auf die Idee, einen Venenstauer neu zu entwickeln? Endlich wurde das Hilfsmittel im medizinischen Alltag neu erdacht und in vielen funktionalen sowie ergonomischen Punkten optimiert. Auch die Hygiene wird dabei auf ein neues Niveau gehoben.

When you stop to think about it, it's actually surprising that nobody has come up with the idea of redesigning the tourniquet before. Now, however, this everyday medical utensil has finally been revisited and many of its functional and ergonomic aspects optimised in the process. That goes for hygiene too, which has been taken to a new level.

HERSTELLER/MANUFACTURER
Kimetec GmbH
Ditzingen

DESIGN
UP Designstudio GmbH & Co. KG
Stuttgart

VERTRIEB/DISTRIBUTOR
Kimetec GmbH
Ditzingen

Oft sind es die kleinen Dinge, die zur besseren Hygiene im Medizinbereich beitragen. Das gilt auch für den Venenstauer, ein für die Blutabnahme unverzichtbares, aber unscheinbares Utensil. Die bislang meist aus textilen Materialien bestehenden Staubänder waren schwer zu reinigen und mussten früh entsorgt werden. Die Verwendung eines Silkonbandes erlaubt nun, das Tool per schneller Wischdesinfektion oder im Autoklaven bei einer Temperatur von 134 Grad Celsius zu sterilisieren. Die feine Musterstruktur im Band verhindert das Abrutschen bei der Anwendung, die Schließe ist ergonomisch geformt, für die Einhandbedienung optimiert und vermeidet Hautquetschungen. Ist der Venenstauer nicht in Gebrauch, lässt er sich dank seines Clips aufhängen oder aufgerollt fixieren. Schließe und Band eignen sich zudem für individuelles und langlebiges Branding.

It's often the little things that help improve hygiene in the healthcare industry. That goes for the tourniquet too, an inconspicuous but nevertheless indispensable aid for taking blood. Because the straps have mostly been made of textile materials up until now, they were difficult to clean and had to be disposed of after just a short time. Thanks to the use of a silicone strap, however, the tool can now either be wiped with disinfectant or sterilised in an autoclave at a temperature of 134 degrees Celsius. The subtle textured pattern prevents the strap from slipping on the patient's arm, while the ergonomically shaped clasp has been optimised for one-handed usage and doesn't bruise the skin. And when it's not in use, the tourniquet can either be hung up by its clip or rolled up and secured. In addition, both the clasp and band are suitable for individual and durable branding.

»Im Bereich Markenauthentizität existieren große Potenziale für Marken und Produkte: gekonnt die Story erzählen und so die emotionale Verbindung zum Nutzer stärken. Als Brandingexperte sehe ich hier noch große Potenziale, auch bei Investitionsgütern.«

»Authenticity offers tremendous potential for brands and products: a story well told forges an emotional connection with the consumer. As a branding expert I am seeing great opportunity, not just for consumers but for B2B as well.«

Roland de Fries, gebürtiger Rheinländer, studierte und arbeitete in Stuttgart, bevor er 2000 nach New York City zog, um bei Ogilvy als Design Director tätig zu sein. Später gründete er die Brandingagentur Hudson Vandam, um speziell deutsche und europäische Kunden im amerikanischen Markt erfolgreich zu positionieren.

Roland de Fries grew up in the Rhineland region and studied and worked in Stuttgart before moving to New York City in 2000. After a period as design director with Ogilvy, he launched his own branding agency for European and German clients in the American market.

www.hudsonvandam.com

www.hudsonvandam.com

1 → SEITE/PAGE
56, 57

BAD, WELLNESS
BATHROOM, WELLNESS

SILVER:
1 **CRADLE-TO-CRADLE CERTIFIED**
Grohe AG
Düsseldorf

Schon lange sind die Zeiten vorbei, als das Badezimmer ein unkomfortabler Raum für die Körperreinigung war. Heute ist das Bad ein emotional und sinnlich aufgeladenes Wellness-Refugium, das höchsten Anforderungen nach Individualität, Ästhetik und Atmosphäre gerecht wird. Besonders die Welt der Armaturen und Keramiken erstaunt durch ihre ästhetische und funktionale Vielfalt.

The days when the bathroom was an uncomfortable space reserved for personal hygiene are well and truly over. Today the bathroom is an emotionally appealing and sensuous wellness retreat that meets the very highest standards in terms of individuality, aesthetics and atmosphere. The world of bathroom fittings and ceramics is particularly impressive thanks to its aesthetic and functional diversity.

SILVER CRADLE-TO-CRADLE CERTIFIED → SEITE/PAGE 57

BAD-ARMATUREN
BATHROOM FITTINGS

SILVER | **CRADLE-TO-CRADLE CERTIFIED** | **BAD-ARMATUREN / BATHROOM FITTINGS**

JURY STATEMENT

Das Design ist weder neu noch prätentiös. Doch mit der Optimierung nach dem C2C-Standard werden die Produkte mit einer neuen, zukunftsgerechten Qualität aufgeladen. Weil in großen Stückzahlen produziert, ist der Nachhaltigkeitseffekt groß und kann auf vielen Ebenen wirken.

The design is not new, nor is it pretentious. But following their optimisation to meet the C2C standard, the products have nevertheless been enhanced with a new, future-friendly quality. And because they're produced in large quantities, the sustainability benefits are considerable and can take effect at many different levels.

HERSTELLER/MANUFACTURER
Grohe AG
Düsseldorf

DESIGN
Lixil Global Design
Düsseldorf

VERTRIEB/DISTRIBUTOR
Grohe Deutschland Vertriebs GmbH
Porta Westfalica

Der Verbrauch an Ressourcen ist eines der großen Probleme des linearen Wirtschaftsmodells. Einen Gegenentwurf dazu stellt das Cradle-to-Cradle-Prinzip dar, bei dem Materialien im beständigen Kreislauf bleiben. Grohe hat nun vier seiner meistverkauften Produkte – drei Armaturen sowie das Brauseset Tempesta 100 – entsprechend modifiziert und sie nach dem Gold-Standard von Cradle-to-Cradle zertifizieren lassen. Dabei wurden neben internen Produktionsabläufen auch vorgelagerte Prozesse optimiert. Zwei der Armaturen nutzen zudem eine veränderte Mischeinheit, die bei Mittelstellung des Einhandhebels lediglich kaltes Wasser liefert.

Um den Kreislaufgedanken weiter zu forcieren, arbeitet das Unternehmen an einem schlüssigen Rücknahmesystem ausgedienter Armaturen.

The consumption of resources is one of the major problems associated with the linear business model. The cradle-to-cradle principle, whereby materials remain part of a continuous cycle, poses an alternative. Now Grohe has modified four of its bestselling products – three taps and the Tempesta 100 shower rail set – and had them certified as meeting the »Gold« standard of the Cradle to Cradle programme. That involved the optimisation not just of internal production methods but of upstream processes as well. In addition, two of the taps use an adapted mixer unit that only supplies cold water when the lever is in the middle position.

What's more, the company is working on a coherent returns system for decommissioned fittings to expedite adoption of the circular principle.

1 → SEITE/PAGE
60-65

2 → SEITE/PAGE
66, 70

3 → SEITE/PAGE
67, 71

4 → SEITE/PAGE
68, 72

KÜCHE, HAUSHALT, TISCHKULTUR
KITCHEN, HOUSEHOLD, TABLE

GOLD:
1 PWM 514, 520, 912, 916, 920
Miele & Cie. KG
Gütersloh

SILVER:
2 PDR 514, 518, 522, 914, 918, 922, 944
Miele & Cie. KG
Gütersloh

SPECIAL MENTION:
3 TRIFLEX HX2
Miele & Cie. KG
Gütersloh

4 SMARTEGG
R3
Stuttgart

Die Technisierung des Haushalts schreitet weiter voran, dank neuer Optionen der Automatisierung und dem Internet of Things. Dabei brillieren hier Produkte, die technologisch höchst durchdacht und nutzungsorientiert konzipiert sowie gestaltet sind. Design sorgt hier für intuitive Zugänglichkeit, bringt innovative Ideen ein und lässt das Versprechen, weniger Zeit für die Hausarbeit aufwenden zu müssen, Wirklichkeit werden.

The technicisation of the household is progressing swiftly, driven by the new options made possible by automation and the internet of things. Products that combine a user-oriented concept and design with highly sophisticated technology shine the brightest. In such cases, design ensures intuitive accessibility, contributes innovative ideas and turns the promise of having to spend less time on household chores into reality.

GOLD | PWM 514, 520, 912, 916, 920 | GEWERBLICHE WASCHMASCHINEN
COMMERCIAL WASHING MACHINES

PWM 514, 520, 912, 916, 920 WASC

KÜCHE, HAUSHALT, TISCHKULTUR
KITCHEN, HOUSEHOLD, TABLE

FOCUS GOLD

GEWERBLICHE WASCHMASCHINEN

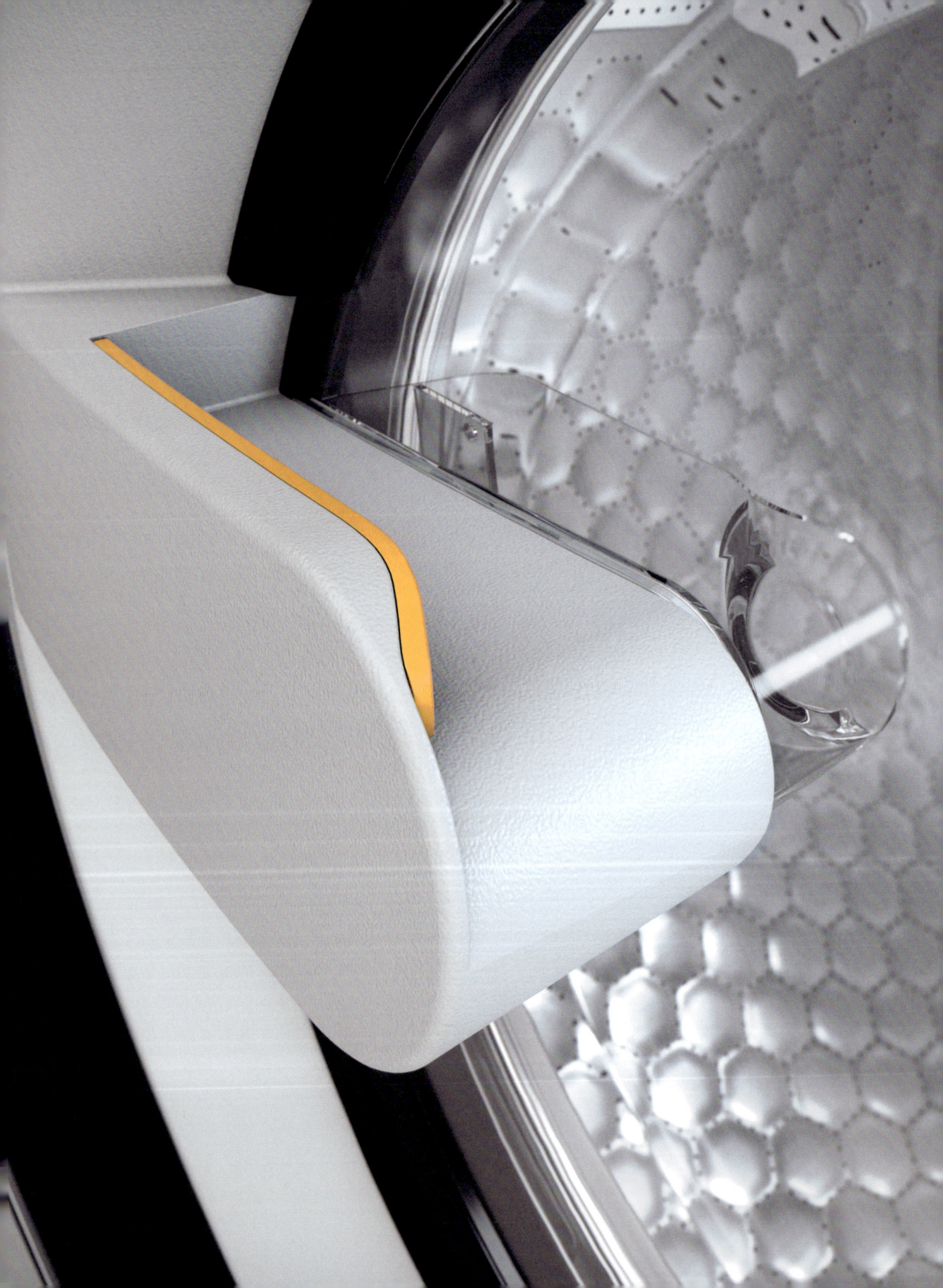

GOLD — PWM 514, 520, 912, 916, 920 — GEWERBLICHE WASCHMASCHINEN / COMMERCIAL WASHING MACHINES

JURY STATEMENT

Obwohl es sich um Industriemaschinen handelt, wurde hier nicht nur auf maximale Funktionalität, sondern auch auf Ästhetik größter Wert gelegt. Auf den schnellen Blick nicht vorhanden, zeigt sich die Gestaltungsqualität in Details und der konsequenten formalen Reduktion. Besonders die Usability macht die Maschinen so bestechend, dazu gehören große Piktogramme ebenso wie die Ausgabe der Bedienoptionen und Anweisungen in Klartext.

While the machines are intended for commercial use, great attention has been paid not just to maximum functionality but to aesthetic aspects as well. Although not immediately obvious, the quality of the design reveals itself in the details and consistently pared-down forms. Usability aspects such as large pictograms and the use of plain text for the setting options and instructions make the machines particularly impressive.

HERSTELLER/MANUFACTURER
Miele & Cie. KG
Gütersloh

DESIGN
Inhouse

VERTRIEB/DISTRIBUTOR
Miele & Cie. KG
Gütersloh

Mit 12 bis 20 Kilogramm Fassungsvermögen sind die neuen Waschmaschinen der PWM-Reihe für den professionellen Einsatz gedacht. Die Maschinen sind formal absolut reduziert und bilden mit ihrer Edelstahl- und Grau-Optik einen selbstbewussten Ruhepol im hektischen Wäschereialltag. Im Vordergrund der Entwicklung stand auch die einfache Bedienbarkeit per Touchdisplay, dessen Klartext-Anzeige in mehreren Sprachen wählbar ist. Durch die Personalisierung der Steuerung und die farbliche Kennzeichnung der abrufbaren Programme eignen sich die Maschinen auch für die Nutzung durch Mitarbeiter:innen mit eingeschränkten kognitiven Fähigkeiten. Ein großer Türgriff rundet das neue Bedienkonzept ab.

Kurze Laufzeiten und geringe Verbräuche sind ebenso wie Langlebigkeit und der weitgehende Verzicht auf Kunststoffe herausragende konstruktive Merkmale. Die Maschinen sind per LAN oder WLAN vernetzbar.

With a load capacity of 12 to 20 kilograms, the new washing machines from the PWM series are intended for professional use. With a form that has been pared down to the absolute minimum and their stainless steel and grey housings, the machines radiate a sense of self-confident calm amid the hectic activity in the laundry room. Development also focused on ease of use, resulting in a touchscreen interface with a plain text display that can be selected in various languages. Thanks to the customisable controls and the colour coding of the preset programmes, the machines are also suitable for use by people with limited cognitive abilities. A large door handle rounds off the new usability concept.

Short cycle times, low water and energy consumption, durability and the almost complete absence of plastics are among the design's most outstanding features. The machines also come with network connectivity via LAN or wifi.

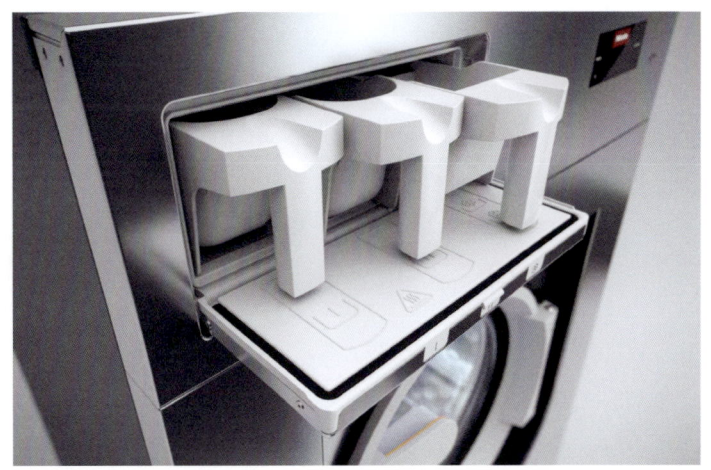

MARKUS HAIN UND HOLGER DE BOER **DESIGN CENTER, MIELE & CIE. KG**

»Miele forciert eine reduzierte, fokussierte Designsprache.«

»Miele is pushing ahead with a reductive, focused design language.«

MARKUS HAIN AND HOLGER DE BOER

DESIGN CENTER, MIELE & CIE. KG

→ **Worin unterscheiden sich gewerbliche und private Waschmaschinen eigentlich?**
In vielen Punkten. Da wären zunächst die größeren Dimensionen, weil hier ganz andere Wäschemengen anfallen. Daraus ergeben sich deutlich höhere Beschleunigungskräfte, die nur eine mitschwingende Trommel auffangen kann. Auch haben wir es mit sehr unterschiedlichen Arten von Wäsche zu tun, beispielsweise von der Pferdedecke bis zur Kleidung von Rettungsdiensten. Das erfordert abgestimmte Spezialprogramme und eine angepasste Verfahrenstechnik. Die Geräte sind für Dreischichtbetriebe ausgelegt, entsprechend robust müssen sie sein.

Sie sprechen davon, dass sich die Steuerung personalisieren lässt. Was ist damit gemeint?
Unsere Maschinen arbeiten in unterschiedlichsten Umgebungen, entsprechend variieren auch die Anforderungen an die User-Interfaces. Während beispielsweise in Großwäschereien oft ausgebildete Spezialist:innen die Bedienung übernehmen, finden wir in Waschsalons ein völlig anderes Nutzungsprofil vor. Hier müssen die Bedienungsoptionen auf das Wesentliche reduziert und intuitiv zugänglich sein. Wir bieten daher weitreichende Optionen zur Konfiguration der Bedienoberfläche, sodass User:innen nur das sehen, was sie wirklich benötigen. Wir können so auch tiefer liegende Expertenfeatures gegen Fehlbedienung absichern.

Die Maschinen erscheinen wie nüchterne Kuben – wo verbirgt sich das Design?
Miele forciert eine reduzierte, fokussierte Designsprache. Die Geräte sind sehr lange im Einsatz, meist über 20 Jahre. Unser Design ist daher von Zeitlosigkeit und Kontinuität geprägt. Letzteres ist wichtig, da die Maschinen oft in Reihe aufgestellt werden und sich einzelne, ausgetauschte Exemplare optisch einfügen sollen. Aber natürlich finden Sie gezielt eingesetzte Formelemente. Die Tür – wie aus einem Guss – inszeniert das robuste Scharnier und den Griffbereich. Die flexibel aufgehängte Trommel verändert ihre Position je nach Beladungsmenge und damit auch die Türposition, was zu einem unruhigen, asymmetrischen Eindruck führen könnte. Der schildförmige Vorderwandausschnitt fängt die variablen Trommelpositionen visuell auf und signalisiert zudem Zuverlässigkeit.

Kann Design die Reparierfähigkeit verbessern?
Für das Design eine besondere Herausforderung. Aber ja, absolut. In Reihenaufstellung sollten alle wichtigen Wartungszugänge von vorne zugänglich sein. Aus Sicherheitsgründen sind meist werkzeuggebundene Öffnungs-Mechanismen verlangt, daher müssen wir in Einzelfällen auf sichtbare Schrauben zurückgreifen. Können wir die Schrauben nicht verstecken, platzieren wir sie so, dass sie die Formensprache unterstützen.

> Von einer 1899 nahe Gütersloh gegründeten Zentrifugierfabrik mit elf Mitarbeitern hat sich die Miele & Cie. KG bis heute zu einem weltweit agierenden Premiumanbieter für Haushaltsgeräte entwickelt. Die Miele-Gruppe beschäftigt weltweit etwa 20.500 Menschen, davon etwa 11.000 in Deutschland.
>
> www.miele.de

→ **How do commercial and private washing machines actually differ?**
In many respects. The bigger dimensions for one thing, because the amounts of laundry that have to be done are on a totally different scale. That results in much stronger accelerative forces that can only be absorbed by a flexibly suspended drum unit. And they have to cope with very different kinds of laundry too: it could be anything from horse blankets all the way to emergency services clothing. That calls for corresponding special programmes and technology that's specially adapted to the processes required. And the appliances are designed for three-shift laundries, so they have to be sufficiently robust.

When you say the controls are customisable, what exactly do you mean?
Our machines are used in all sorts of different settings, and the requirements the user interfaces have to meet vary accordingly. If they're deployed in a commercial laundry business they might be operated by trained specialists, whereas the usage profile in launderettes is totally different. In the latter case, the control options have to be reduced to the absolute essentials and be intuitive to use. That's why we offer extensive options for configuring the interface so that users only see what they really need. That also enables us to protect deep-level expert features from incorrect use.

The machines look like plain cubes – where exactly does design come into play?
Miele is pushing ahead with a reductive, focused design language. The appliances are in use for a very long time, usually more than 20 years. That's why timelessness and continuity are the main considerations. Continuity is important because the machines are often set up in rows, so if an individual appliance has to be replaced the new one should blend in. But obviously there are some very deliberate distinctive elements too. The door is designed to look like a single, seamless part and shows off the robust hinge and the handle area. Because of its flexible suspension, the drum changes position depending on the size of the load, causing the door to do the same – which could potentially make an inharmonious, asymmetric impression. That's why the design features a shield-shaped cutout in the front: at a visual level, it offsets the variable drum position and also signals reliability.

Can design improve repairability?
That's particularly challenging. But yes, absolutely. When the machines are set up in rows, it should be possible to get to all the important maintenance access points from the front. For safety reasons, tool-dependent opening mechanisms are usually required, so now and again we have to resort to visible screws. If we can't conceal the screws, we position them in such a way that they underscore the design language.

> Founded near Gütersloh as a centrifuge factory with 11 employees in 1899, Miele & Cie. KG has evolved into a globally active supplier of premium household appliances. The Miele Group employs approx. 20,500 people all over the world, around 11,000 of them in Germany.
>
> www.miele.de

SILVER   PDR 514, 518, 522, 914,   GEWERBLICHE WÄSCHETROCKNER
         918, 922, 944   COMMERCIAL TUMBLE DRYERS
         → SEITE/PAGE
         70

| SPECIAL MENTION | TRIFLEX HX2　→ SEITE/PAGE 71 | AKKU-STAUBSAUGER CORDLESS VACUUM CLEANER |

SPECIAL MENTION | SMARTEGG → SEITE/PAGE 72 | IONISATOR IONISER

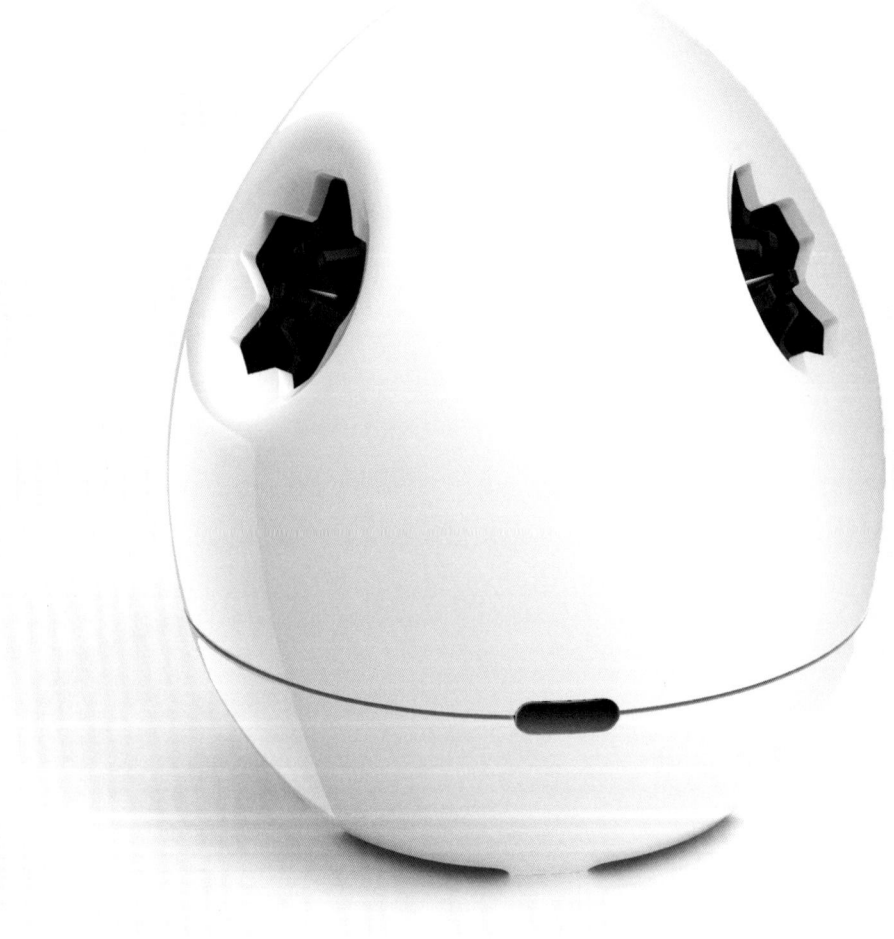

SILVER

PDR 514, 518, 522, 914, 918, 922, 944

**GEWERBLICHE WÄSCHETROCKNER
COMMERCIAL TUMBLE DRYERS**

JURY STATEMENT

Sehr klar gestaltete Maschine, in purer geometrischer Form belassen. Die Trockner machen nicht mehr aus sich als sie funktional sein müssen. Die schlichte, graue Erscheinung wird an entscheidenden Stellen aufgebrochen, etwa in Form des Notaus-Tasters, des Displays oder des gelb akzentuierten Türöffners.

A machine that owes the clarity of its design to its pure, geometric form. The dryers don't make any more of themselves than their function requires. Their simple grey appearance is interrupted only in crucial places, by features such as the emergency stop button, the display and the yellow strip that accentuates the door opener.

HERSTELLER/MANUFACTURER
Miele & Cie. KG
Gütersloh

DESIGN
Inhouse

VERTRIEB/DISTRIBUTOR
Miele & Cie. KG
Gütersloh

Mit den großvolumigen Trocknern ergänzt Miele sein Portfolio für den industriellen Einsatz. Trotz ihres unterschiedlichen Fassungsvermögens zwischen 10 und 44 Kilogramm sind die Maschinen von einem durchgängigen Designprinzip geprägt, das strenge formale Reduktion pflegt und das Werkzeug als solches belässt. Viel Energie steckte das Designteam in die nutzerzentrierte und intuitive Bedienung. Vier Steuerungsvarianten, drei mit Drehwahlschalter und eine mit Sensortouch-Technik erlauben den direkten Zugriff auf die zentralen Funktionen. Edelstahl und Eisengrau prägen die farbliche Präsenz der sehr langlebigen Anlagen, deren Konstruktion auch die spätere Recyclingfähigkeit unterstützt.

Miele is extending its portfolio of appliances for industrial use with these large-volume dryers. Despite different load capacities ranging between 10 and 44 kilograms, the machines are based on a consistent design principle that reduces forms to the minimum and lets the tool be a tool. The design team invested a great deal of energy in the user-centric and intuitive interface. The controls come in four different variants – three with a dial and one with SensorTouch technology – that enable the user to access the key functions directly. Available with a stainless steel or iron grey housing, the appliances are not only extremely durable, their future recyclability has been factored into the design as well

SPECIAL MENTION · TRIFLEX HX2 · AKKU-STAUBSAUGER / CORDLESS VACUUM CLEANER

JURY STATEMENT

Ein ausgesprochen praktischer Helfer im Haushalt, weil seine Konvertierbarkeit weitere staubsammelnde Geräte unnötig macht. Der im Kern technische Look unterstreicht den Werkzeugcharakter, die neue Motorisierung ist eine sehr sinnvolle Produktpflege.

An extremely practical household helper because its convertibility makes other dust-collecting appliances superfluous. Its essentially technical look underscores the fact that this is a tool, while the new motor is a very meaningful product update.

HERSTELLER/MANUFACTURER
Miele & Cie. KG
Gütersloh

DESIGN
Inhouse

VERTRIEB/DISTRIBUTOR
Miele & Cie. KG
Gütersloh

Bereits das Vorgängermodell dieses kabellosen Handstaubsaugers überzeugte sowohl durch seine multiflexiblen Umbauoptionen als auch durch eine betont zurückhaltende Formensprache. Bei der Konzeption der aktuellen Serie rückte das Design nochmals in den Fokus – Formmerkmale, Farben und Metallakzente folgen nun einem übergeordneten Styleguide des Herstellers. Neben verbesserter Ergonomie und Bedienfreundlichkeit sorgt vor allem der leistungsstärkere Motor im Zusammenspiel mit der intelligenten Bodenbelagserkennung der Bürstenwalze für längere Akkulaufzeiten und damit für mehr Komfort beim Saugeinsatz.

The predecessor model of this battery-powered handstick vacuum cleaner was already thoroughly convincing thanks to both its ultra-flexible configuration options and its emphatically understated design language. The concept behind the latest series focused on the design – the form-based features, colours and metal accents now conform to the manufacturer's overarching style guide. Besides the improved ergonomics and ease of handling, the combination of a more powerful motor and smart floorhead ensures extended battery life and thus greater convenience for the user.

SPECIAL MENTION — SMARTEGG IONISATOR / IONISER

JURY STATEMENT

Interessant, einem rein technischen Prozess eine so natürliche und harmonische Form zu verleihen. Auf den zweiten Blick entsteht durch die offensichtliche Integration der Öffnungen und einer Fuge ein spannendes Objekt mit vielversprechender Funktion.

The choice of such a natural, harmonious shape for a purely technical process is an interesting approach. On closer inspection, the obvious integration of the apertures and a seam results in an intriguing object with a promising function.

HERSTELLER/MANUFACTURER
R3
Stuttgart

DESIGN
Red GmbH
Stuttgart

VERTRIEB/DISTRIBUTOR
R3
Stuttgart

Ionen sind positiv oder negativ geladene Atome und Moleküle, die aufgrund ihres elektrischen Potenzials in der Lage sind, kleinste Partikel wie Staub, Pollen oder Pilzsporen anzuziehen und sich mit ihnen zu verbinden. Sie werden deshalb oft in Luftreinigungsgeräten, aber auch in der Lebensmittelindustrie eingesetzt, um die Belastung von Nahrungsmitteln mit Bakterien und Schimmelsporen zu reduzieren.

Der wieder aufladbare, eiförmige Ionisator ist dazu gedacht, seinen Dienst entweder im Kühl- und Kleiderschrank privater Haushalte oder am Arbeitsplatz zu verrichten. Über drei Düsen werden die in seinem Inneren erzeugten Ionen nach außen abgegeben, wo sie Viren, Bakterien, Sporen und Gerüche bekämpfen. Etwa ein Monat darf vergehen, bevor der Ionisator über einen USB-C-Port wieder an die Steckdose muss.

Ions are positively or negatively charged atoms and molecules; because of their electric potential, they are able to attract tiny particles such as dust, pollen or fungal spores and combine with them. As a result they are often used in air purifiers, but also in the food industry, where they reduce contamination with bacteria and mould spores.

The rechargeable egg-shaped ioniser is intended for use in refrigerators and wardrobes in private households or at the workplace. Via three openings, the ions produced inside it are released into the surrounding space, where they combat viruses, bacteria, spores and odours. The ioniser can remain in place for about a month without needing to be hooked up to a power outlet via a USB-C port.

»In der Maschinenbau-Branche gibt es kaum ein Produkt ohne Software, das Zusammenspiel von Hard- und Software ist hier besonders wichtig, was mitunter eine besondere Herausforderung darstellt.«

»In the mechanical engineering sector it's hard to find a product that doesn't rely on software: it's a field where the interaction between hard- and software is particularly important, and that can sometimes be extremely challenging.«

Dina Gallo studierte Industriedesign an der Staatlichen Akademie der Bildenden Künste in Stuttgart. Sie war unter anderem Leiterin des Designmanagements bei Hansa Metallwerke, absolvierte parallel dazu ein MBA Studium an der Steinbeis-Hochschule Berlin und war Lehrbeauftragte an der HfG Schwäbisch Gmünd. Seit 2012 verantwortet sie das Designmanagement der Trumpf-Gruppe.

Dina Gallo studied industrial design at Stuttgart State Academy of Art and Design (ABK Stuttgart). She went on to become head of design management at Hansa Metallwerke, during which time she also did an MBA at Steinbeis University Berlin, and was an associate lecturer at the University of Design (HfG) Schwäbisch Gmünd. She has been head of design management for the Trumpf Group since 2012.

www.trumpf.com

www.trumpf.com

1 → SEITE/PAGE
78, 84

2 → SEITE/PAGE
79, 85

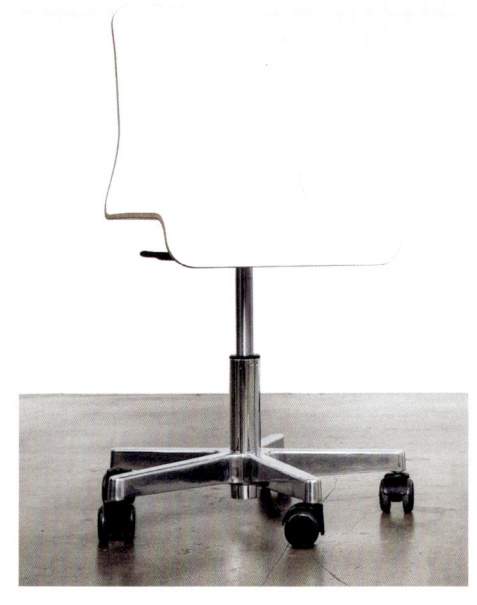

3 → SEITE/PAGE
80, 86

4 → SEITE/PAGE
81, 87

5 → SEITE/PAGE
82, 88

INTERIOR
INTERIORS

SILVER:
1 **BRIDGE**
Mobimex AG
Seon
Schweiz/Switzerland

2 **U-TURN**
Bene GmbH
Waidhofen an der Ybbs
Österreich/Austria

SPECIAL MENTION:
3 **STUDDY**
Pure Position by IWL gGmbH
Machtlfing

4 **MOVE 75 ELEGANCE**
Gross + Froelich GmbH & Co. KG
Weil der Stadt

5 **X-COMPANIONS**
wd3 GmbH
Stuttgart

Ein Stuhl, ein Tisch, ein Bett und ein Regal – was braucht man mehr? Und doch ist das Universum der Möbel immens weit, wandelt sich stetig, erneuert sich und spielt mit Volumina, Materialien, Oberflächen, Farben. Möbeldesign ist eine der populärsten Gestaltungsdisziplinen, die gerade in Zeiten des Homeoffice ganz neu durch die Prämissen Flexibilität und Funktionalität gefordert wird.

A chair, a table, a bed and a shelving unit – what more does anyone need? And yet the world of furniture is truly immense, changes constantly, reinvents itself and plays with volumes, materials, finishes and colours. Furniture design is one of the most popular design disciplines of all – and in home office times like these, it is facing totally new challenges in terms of flexibility and functionality.

SILVER BRIDGE TISCH
→ SEITE/PAGE TABLE
84

SILVER U-TURN WORKSHOP-SITZ
               WORKSHOP SEAT

→ SEITE/PAGE
  85

| SPECIAL MENTION | STUDDY → SEITE/PAGE 86 | DREHSTUHL SWIVEL CHAIR |

SPECIAL MENTION

MOVE 75 ELEGANCE
→ SEITE/PAGE 87

MÖBELROLLE
FURNITURE CASTER

SPECIAL MENTION	X-COMPANIONS → SEITE/PAGE 88	MULTIFUNKTIONSMÖBEL MULTIFUNCTIONAL FURNITURE

SILVER

BRIDGE

TISCH
TABLE

JURY STATEMENT

Ein sehr gelungener Tischentwurf mit einem faszinierenden Spannungsbogen zwischen der organischen Holzplatte und dem rigiden Betonfuß. Trotz dieser sehr eigenständigen Erscheinung passt der Tisch zu vielen Stuhlmodellen und Interieurs, ist also sehr universell nutzbar.

A very successful table design that owes its fascination to the intriguing combination of organic wooden top and rigid concrete pedestal. Despite its highly original appearance, the table pairs well with a wide variety of chairs and interiors and is therefore almost universally compatible.

HERSTELLER/MANUFACTURER
Mobimex AG
Seon
Schweiz/Switzerland

DESIGN
Jehs & Laub
Stuttgart

VERTRIEB/DISTRIBUTOR
Mobimex AG
Seon
Schweiz/Switzerland

Eine ungewöhnliche Formensprache kennzeichnet dieses Tischkonzept mit seinen eleganten, fließenden Formen. Optisch fast schwerelos, ruht die organisch geformte Tischplatte auf einem konischen Unterbau aus Beton. Größere Platten lagern auf zwei Betonkörpern, wobei die Verbindung nie direkt erfolgt, sondern über ein schlankes, aussteifendes Stahlträgergestell dazwischen. Auch bei weit spannenden Tischen bleibt die Plattenstärke so auf schlanke 28 Millimeter begrenzt, die Kantenfase verstärkt den schwebenden Eindruck. Sowohl für den Wohn- wie den Objektbereich gedacht, also für das Tête-à-Tête ebenso wie für das harte Business-Meeting, entwickelt sich zwischen dem Betonunterbau und den unterschiedlich behandelten Massivholzplatten aus Eiche oder Schwarznuss ein lebendiger Materialkontrast.

The table concept speaks an unusual design language defined by elegant, fluid forms. Almost weightless in appearance, the organically shaped tabletop rests on a conical concrete base. The larger tops are supported by two concrete cones but never directly connected with them; instead, a slender steel frame in between ensures the necessary support. As a result, even tables with a wide span can be topped with a slender 28-millimetre-thick panel, with chamfered edges that reinforce the floating impression. Intended for private settings as well as the contract sector, and therefore both for a cosy tête-à-tête or a tough business meeting, the tables are available with oak or black walnut tops in a choice of finishes, creating a vivid contrast between the expanse of solid wood and the concrete base that supports it.

SILVER — U-TURN — WORKSHOP-SITZ / WORKSHOP SEAT

JURY STATEMENT

Der überraschend bequeme Sitz fügt sich dank seiner filigranen Ausarbeitung in ganz unterschiedliche Kontexte ein. Auch die Nutzung zeichnet sich durch hohe Flexibilität und Anpassbarkeit aus. Details und Übergänge der einzelnen Konstruktionselemente sind gut gelöst, Proportionen und Wertigkeit stimmen.

Thanks to its slender design, the surprisingly comfortable seat blends in with a wide variety of contexts and provides a high degree of flexibility and adaptability in terms of usage. Compelling solutions have been found for the details and transitions between the individual elements, the proportions are harmonious and the communication of value appropriate.

HERSTELLER/MANUFACTURER
Bene GmbH
Waidhofen an der Ybbs
Österreich/Austria

DESIGN
Inhouse
Christian Horner

VERTRIEB/DISTRIBUTOR
Bene GmbH
Waidhofen an der Ybbs
Österreich/Austria

Mobil in doppelter Hinsicht macht diese an eine schmale Bank erinnernde Sitzgelegenheit: Sie hält die auf ihr sitzende Person durch die eher stehende Position in Bewegung und lässt sich in der Rollen-Version selbst im Raum umherfahren. Gedacht für agile Workshops oder schnelle Meetings, präsentiert sich die Sitz-Tisch-Kombination bewusst einfach. Unter der Sitzfläche lassen sich optional Utensilien verstauen und mit einem Schloss sichern. Der schwenkbare Aufsatztisch mit drei Kilogramm Belastbarkeit und 50 Zentimetern Breite ist für die Arbeit mit Papierunterlagen oder Notebook konzipiert. Dank der Sitzhöhe von 69 Zentimetern eignet sich U-Turn auch als Unterstützung an höhenverstellbaren Arbeitstischen. Bezüge sind in verschiedenen Stoffen oder auch Leder erhältlich, das Polster darunter besteht aus Polyurethan-Schaumstoff.

Reminiscent of a narrow bench, this seat mobilises in two ways: the user stays in permanent motion due to a position that is almost closer to standing, while the version with casters allows the seat itself to move around the room. Because the product is meant for agile workshops or quick meetings, the design of the seat-and-table combination is deliberately simple. The version with a fold-up seat comes with a storage compartment that can be secured with a lock if required. Intended for working with documents or a laptop, the swivel table attachment is 50 centimetres wide and can support a load of up to 3 kilograms. Thanks to a seat height of 69 centimetres, the U-Turn is also suitable for use with height-adjustable desks. The cover is available in various fabrics or leather, the upholstery is made of polyurethane foam.

SPECIAL MENTION STUDDY DREHSTUHL / SWIVEL CHAIR

JURY STATEMENT

Ein sehr schönes Produkt für den Kindermöbel-Markt, das dank seiner Schlichtheit tatsächlich in der Lage ist, die formalen Präferenzverschiebungen der jungen Nutzer:innen über die Jahre hinweg zu überstehen. Und auch der soziale Hintergrund der Fertigung ist vorbildlich.

A very attractive product for the children's furniture market. Thanks to its simplicity, it is genuinely capable of withstanding the shifts in its young users' design preferences over the years. What's more, the social backstory to its production is exemplary.

HERSTELLER/MANUFACTURER
Pure Position by IWL gGmbH
Machtlfing

DESIGN
Olaf Schroeder Industrial Design
Berlin

VERTRIEB/DISTRIBUTOR
Pure Position by IWL gGmbH
Machtlfing

Speziell auf Kinder und Jugendliche zwischen 6 und 16 Jahren abgestimmt ist dieser Drehstuhl. Sein bewusst schlichtes Design verhindert, dass der Stuhl im Laufe des Heranwachsens nicht mehr altersgerecht erscheint und ausgetauscht wird. Die Sitzschale aus HPL-beschichtetem Multiplex-Formholz zeigt eine dynamische Linienführung und ist ausgesprochen langlebig. Um die Bewegungsfreiheit des Kindes nicht einzuschränken, wird auf Armlehnen verzichtet. Das Konzept der Einfachheit spiegelt sich auch in der Tatsache wider, dass der Stuhl per Gasdruckfeder lediglich in der Höhe verstellbar ist.

Der Stuhl ist Teil einer kompletten Kindermöbel-Serie und wird von den bayerischen IWL-Werkstätten von Menschen mit Behinderung produziert und vertrieben.

This swivel chair is specifically geared towards children and teenagers between the ages of 6 and 16. Its deliberately simple design prevents the chair from seeming too babyish as its young user grows up and being replaced as a result. Made of HPL-coated multiplex plywood, the seat shell is defined by dynamic lines and is extremely durable. There are no armrests so as not to restrict the child's freedom of movement. The simplicity principle is likewise reflected in the fact that there is only one adjustment option: the height of the gas lift seat.

The chair is part of a complete series of children's furniture and is produced and distributed by IWL-Werkstätten, a Bavarian nonprofit organisation that employs people with disabilities in its workshops.

SPECIAL MENTION

MOVE 75 ELEGANCE
MÖBELROLLE / FURNITURE CASTER

JURY STATEMENT

Obwohl konstruktiv für schwere Lasten ausgelegt, ist dies der Rolle nicht anzusehen. Sie ist schlank und formal klar gestaltet und eignet sich daher ideal als Zukaufelement für engagiertes Möbeldesign, das sich auch im Wohnbereich nicht verstecken muss.

Although the caster is designed for heavy loads, you wouldn't know by looking at it. Thanks to its clear and slender forms, the product is an ideal bought-in component for stylish furniture design and looks equally at home in a commercial or domestic setting.

HERSTELLER/MANUFACTURER
Gross + Froelich GmbH & Co. KG
Weil der Stadt

DESIGN
Inhouse

VERTRIEB/DISTRIBUTOR
Gross + Froelich GmbH & Co. KG
Weil der Stadt

Das aktuelle Möbeldesign tendiert dazu, Rollen mit größeren Durchmessern einzusetzen. Diesen Trend nimmt die in Deutschland gefertigte Möbelrolle mit ihrem Raddurchmesser von 75 Millimetern auf und kombiniert diese Eigenschaft mit einer hohen Belastbarkeit von 75 Kilogramm pro Rolle. Obwohl funktional eine Schwerlastrolle, spiegelt sich dies in der optischen Erscheinung nicht wider – das Gehäuse aus matt- oder glanzverchromtem Zinkdruckguss bleibt, wie auch die Rolle selbst, schmal und filigran. Auf minimalen Verschleiß ausgelegt, lässt sich die Rolle am Ende ihrer Nutzungszeit materialgerecht demontieren.

Als Weiterentwicklung eines bewährten Rollentyps eignet sich das Produkt zur Ausstattung von Arbeitsstühlen, für Polstermöbel, für Regale oder anderes Mobiliar, das seinen Standort so flexibel wechseln kann. Die Rolle wird freilaufend, gebremst, mit Feststeller und für verschiedene Bodenarten angeboten.

Contemporary furniture design tends to favour casters with larger diameters. The made-in-Germany product responds to this trend with a diameter of 75 millimetres, combined with a load capacity of 75 kilograms per caster. Although a heavy-duty caster in terms of its functionality, this is not reflected in its appearance – both the die-cast zinc housing with a matt or shiny chrome finish and the wheel itself are slender and elegant. Designed for minimal wear and tear, the caster can be disassembled into mono-material components when it reaches the end of its useful life.

The product is an evolution of an established caster family and suitable for use with office chairs, upholstered furniture, shelving or other furnishings intended for flexible deployment. The caster is available in freewheel and locking versions, with a braking mechanism and for various types of flooring.

SPECIAL MENTION

X-COMPANIONS

MULTIFUNKTIONSMÖBEL
MULTIFUNCTIONAL FURNITURE

JURY STATEMENT

Dass gerade die Simplifizierung der Dinge große Chancen zur stetigen Erweiterung der Nutzungsszenarien bietet, beweist dieses System beispielhaft. Zusatzelemente ergänzen die Grundidee und damit die Anwendung. Dabei ist alles intuitiv anwendbar und damit enorm flexibel.

As this system demonstrates in exemplary fashion, simplification is often the key that unlocks the greatest potential for expanding usage scenarios. The add-on elements increase the versatility of the basic idea and thus also the range of applications. At the same time, the products are highly intuitive and therefore extremely flexible.

HERSTELLER/MANUFACTURER
wd3 GmbH
Stuttgart

DESIGN
Inhouse

VERTRIEB/DISTRIBUTOR
wd3 GmbH
Stuttgart

Vor vier Jahren fing alles an: Der Xbrick, ein multifunktionales Sitzelement für die dynamische Raumnutzung, überzeugte nicht nur durch seine fast unbegrenzte Konfigurierbarkeit, sondern auch durch die Fertigung in expandiertem Polypropylen, was den Hocker komplett recycelbar macht.

Modular gedacht war der Xbrick damals schon, nun werden seine Einsatzmöglichkeiten noch einmal umfangreich erweitert. So lässt das Holzelement X-tseat aus dem Xbrick eine Sitz-Tisch-Kombination entstehen, mit X-bench werden zwei Hocker zu einer Sitzbank. X-table eröffnet die Möglichkeit, einen stabilen Tisch zu bauen, während die X-toolbox Utensilien aufnimmt. Mittels Steck- und Gurtverbindungen klappt der Umbau problemlos; und es können immer wieder neue Umgebungen entstehen.

It all started four years ago: the Xbrick, a multifunctional seating element for making dynamic use of the space available, was impressive not just for its almost unlimited configurability but also because the expanded polypropylene it's made of means the stool is completely recyclable.

Although the Xbrick was based on a modular concept right from the start, its range of applications has been considerably expanded by the latest additions. The X-tseat, for instance, is a wooden element that can be used to turn the Xbrick into a seat-and-table combination, while the X-bench transforms two of the stools into a seat for several people. X-table provides the option of building a sturdy table, and the X-toolbox provides practical storage for utensils. Because the individual elements are held together by plug-in connectors and belts, reconfiguring them to create new environments is no problem at all.

1 → SEITE/PAGE
   92, 94

2 → SEITE/PAGE
   93, 95

LIFESTYLE, ACCESSOIRES
LIFESTYLE, ACCESSORIES

SPECIAL MENTION:
1 **PICA FINE DRY**
Pica-Marker GmbH
Kirchehrenbach

2 **MONDO GMT**
Botta Design
Königstein

Sich mit positiven, sinnlich anregenden oder funktional durchdachten Dingen zu umgeben, macht den Alltag einfacher, facettenreich und inspirierend. Das gilt insbesondere für die kleinen Produkte, die uns durch den ganzen Tag begleiten, uns in bestimmten Situationen unterstützen, das Leben erleichtern oder ganz einfach zur Freude gereichen.

Surrounding oneself with positive things that appeal to the senses or are equipped with clever functions makes everyday life easier, more varied and more inspiring. That particularly applies to little products that accompany us throughout the day, provide support in certain situations, make life easier or simply give us pleasure.

SPECIAL MENTION | PICA FINE DRY
→ SEITE/PAGE 94

SCHREIBGERÄT
AUTOMATIC PENCIL

SPECIAL MENTION — MONDO GMT → SEITE/PAGE 95 — ARMBANDUHR / WRISTWATCH

SPECIAL MENTION | PICA FINE DRY | SCHREIBGERÄT / AUTOMATIC PENCIL

JURY STATEMENT

Ein professionell gestaltetes Produkt mit hoher Funktionalität, primär von der Praxis aus gedacht. Dazu gehören zum Beispiel der Schutz der Mine bei Nichtbenutzung sowie die optimierte Griffigkeit des Stiftes. Die Farbigkeit ist ein unverwechselbares Markenmerkmal.

A professionally designed product that delivers a high level of functionality and is primarily based on practical considerations, including the protection of the lead when the pencil is not in use and the optimised grip. The use of colour ensures unmistakable branding.

HERSTELLER/MANUFACTURER
Pica-Marker GmbH
Kirchehrenbach

DESIGN
Winkelbauer Design
Ludwigsburg
und/and
Inhouse
Stephan Möck

VERTRIEB/DISTRIBUTOR
Pica-Marker GmbH
Kirchehrenbach

Beim Markieren im Arbeitsalltag sind belastbare Schreibgeräte mit Präzisionsminen gefragt. Ein äußerst formstabiler Spezial-Kunststoff sorgt für Stabilität und Langlebigkeit des nachfüllbaren Druckbleistifts, der mit zwei unterschiedlichen Griffzonen für das Markieren und Schreiben ausgestattet ist. Im Inneren übernimmt eine Hightech-Druckmechanik den automatischen Minenvorschub im Edelstahlrohr, ein spezieller Minendurchmesser von lediglich 0,9 Millimetern macht das Anspitzen überflüssig. Zwei weitere Pluspunkte im Arbeitseinsatz: Der als Schirmkappe gestaltete Drückerknopf und ein Dichtring am Schaft halten Staub und Nässe ab.

In day-to-day work settings, hardwearing marking tools with high-precision leads are sought-after utensils. An extremely robust special plastic ensures the sturdiness and durability of the refillable automatic pencil, which features two different grip zones for marking and writing. Inside the stainless steel tube, a hi-tech push mechanism operates the automatic lead feed, while a special lead diameter of just 0.9 millimetres makes sharpening superfluous. The pencil comes with two other bonuses for work situations: a pushbutton in the form of an umbrella cap and a sealing ring on the barrel provide protection from dust and moisture.

| SPECIAL MENTION | MONDO GMT | ARMBANDUHR WRISTWATCH | 94 95 |

JURY STATEMENT

Erneut eine Uhr, bei der die Zeitanzeige auf ganz eigene Art interpretiert und angezeigt wird. Die Armbanduhr differenziert sich so auf den ersten Blick am Markt der reinen Zeitmesser. Außerdem überrascht das für die Größe des Gehäuses geringe Gewicht.

A watch that again succeeds in interpreting the art of time display in a highly original way and therefore immediately stands out in the market as compared to other products intended exclusively as chronometers. Given the size of its case, the watch is also surprisingly light.

HERSTELLER/MANUFACTURER
Botta Design
Königstein

DESIGN
Inhouse

VERTRIEB/DISTRIBUTOR
Botta Design
Königstein

In Zeiten digitaler Web-Konferenzen mit Kollegen auf anderen Kontinenten oder Zoom-Meetings mit entfernt lebenden Freunden und Verwandten stellt sich immer wieder die Frage: Wie spät ist es dort? Darauf gibt die kompakte Analoguhr präzise Auskunft. Sie zeigt auf dem Zifferblatt per weißem Stunden- und Minutenzeiger die lokale Zeitzone in 12-Stunden-Skalierung sowie eine weitere, frei wählbare Zeit im 24-Stunden-Format an. Ein künstlicher Horizont gliedert das 24-Stunden-Zifferblatt in eine Tag- und Nachthälfte. Der zugehörige Zeiger mit orangefarbiger Spitze legt innerhalb von 24 Stunden eine Umdrehung zurück.

Das Gehäuse aus Titan ist hautfreundlich, das eingebaute Schweizer Uhrwerk verspricht Langlebigkeit und Präzision.

In the days of virtual meetings with colleagues on other continents and zoom get-togethers with friends and family who live far away, there's one question that comes up again and again: how late is it there? This compact analogue watch provides certainty. Besides indicating the local time zone with white hour and minute hands assigned to a 12-hour scale, the face also shows another, freely selectable time in 24-hour format. This 24-hour scale is divided into daytime and night-time halves by an artificial horizon and orbited once in 24 hours by an orange-tipped hand.

The skin-friendly titanium case contains a Swiss movement that promises both longevity and precision.

JOA HERRENKNECHT

STUDIO JOA HERRENKNECHT, BERLIN/TORONTO

»Die Jurierung war sehr professionell, ich habe viel gelernt und war erstaunt über die Kompetenzen der Unternehmen, die hier dabei sind. Das sind starke Zeichen.«

»The judging was very professional, I learned a great deal and found the expertise of the companies that took part astonishing. They're setting a very strong example.«

Joa Herrenknecht studierte bis 2010 an der HfG Karlsruhe, unter anderem bei Stefan Diez und James Irvine. 2012 gründete sie in Berlin das Studio Joa Herrenknecht und arbeitet multidisziplinär in den Bereichen Produktdesign, Packaging und Interior Design.

Joa Herrenknecht lebt und arbeitet in Berlin und Toronto und gehört zu den Gründerinnen von Matter of Course – einem Designkollektiv von elf Designerinnen mit Sitz in Berlin.

Joa Herrenknecht studied at Karlsruhe University of Arts and Design until 2010, where her teachers included Stefan Diez and James Irvine. She founded Studio Joa Herrenknecht in Berlin in 2012 and brings a multidisciplinary approach to her work in product, packaging and interior design.

Joa Herrenknecht lives and works in Berlin and Toronto and is a co-founder of Matter of Course, an 11-member female design collective based in Berlin.

www.joa-herrenknecht.com

www.joa-herrenknecht.com

1 → SEITE/PAGE 100–105

2 → SEITE/PAGE 106, 107

LICHT
LIGHTING

GOLD:
1 **BEAMER NEW**
Erco GmbH
Lüdenscheid

SPECIAL MENTION:
2 **REFLEX² FLOOR**
Serien Raumleuchten GmbH
Rodgau

Licht aus der Leuchtdiode ist längst Standard – denn sie bietet nicht nur Energieeffizienz, sie ermöglicht auch ganzheitlich konzipierte Leuchtensysteme. Neben der Ergänzung mit zusätzlichen Features oder der faszinierenden Miniaturisierung bietet das Halbleiter-Leuchtmittel die Möglichkeit, auch einzelne Leuchten in digitale Steuerungssysteme zu integrieren.

Light-emitting diodes have long since become a standard light source – in addition to being energy-efficient, they permit holistically designed lighting systems as well. Besides enabling additional features and a fascinating degree of miniaturisation, the semiconductor light source also provides the option of integrating individual luminaires into digital control systems.

GOLD | BEAMER NEW | AUSSEN-SCHEINWERFER OUTDOOR PROJECTOR

BEAMER NEW

AUSSEN— SCHEINW

LICHT
LIGHTING

FOCUS
GOLD

ERFE

GOLD — BEAMER NEW
AUSSEN-SCHEINWERFER / OUTDOOR PROJECTOR

JURY STATEMENT

Obwohl auf Robustheit ausgelegt, besticht das Gehäuse durch seine Schlankheit und seine sehr klare Detaillierung. Das Design fügt sich in verschiedene Architekturen ein, ist aber dennoch eigenständig und prägnant. Gut gelöst wurde die Problematik der Lichtverschmutzung durch die einfache Anpassung der Lichtcharakteristik.

Although designed for robustness, the housing is impressive for its slenderness and very clear detailing. The design blends in with various architectural styles but is nevertheless original and striking. The straightforward adjustability of the light characteristics provides a good solution to the problem of light pollution.

HERSTELLER/MANUFACTURER
Erco GmbH
Lüdenscheid

DESIGN
Inhouse

VERTRIEB/DISTRIBUTOR
Erco GmbH
Lüdenscheid

Die Fassadenbeleuchtung oder die Illuminierung öffentlicher Räume wird unter dem Vorzeichen der Lichtverschmutzung kritisch gesehen. Letztlich kommt es darauf an, das in den Himmel abstrahlende Streulicht zu minimieren oder ganz auszuschließen. Vor diesem Hintergrund wurde dieser Scheinwerfer konzipiert, dessen speziell entwickelte Linsentechnologie einen präzise auf das Objekt abstimmbaren Lichtkegel produziert. Zehn Lichtverteilungen von akzentuierend bis flächig können abgerufen werden. LEDs und Treiberelektronik wurden auf eine Lebensdauer von 100.000 Stunden ausgelegt, ebenso das Kunststoffgehäuse, dessen Gewicht Tragstrukturen an Fassaden wenig belastet.

Die LEDs produzieren Licht mit 3.000 K oder 4.000 K, mittels Konversionsfilter werden weitere acht Lichtspektren erzeugt. Die Steuerung selbst kann auch per Smartphone via Casambi Bluetooth erfolgen, die DALI-Integration ist ebenso machbar.

In the context of light pollution, facade lighting and the illumination of public spaces are seen critically. Ultimately, the goal is to minimise or totally eliminate the light spill that is emitted into the sky. That was the background behind the design of this projector, whose specially developed lens technology produces a beam that can be precisely adjusted to the building or object to be illuminated. Ten distribution options ranging from accentuation to wallwash can be selected. The LEDs and drive electronics are designed for a service life of 100,000 hours, as is the plastic housing – a lightweight solution that minimises the load on the facade mountings.

The LEDs produce light with a colour temperature of 3,000K or 4,000K, and eight other spectra can be produced via conversion filters. The lights can also be controlled with a smartphone via Casambi Bluetooth and are also suitable for integration with a DALI network.

HENK KOSCHE — GROUP MANAGER DESIGN, ERCO GMBH

»Die Herausforderung besteht heute in der Effizienz, Effektivität und smarten Ansteuerung neuer Beleuchtungskonzepte.«

»Today the challenge lies in ensuring the efficiency, effectiveness and smart control of new lighting concepts.«

HENK KOSCHE
GROUP MANAGER DESIGN, ERCO GMBH

Sie nutzen für das Gehäuse Kunststoff – warum keinen Metallguss?
Kunststoffe sind hochwertigste Materialien, die Problemlöser in den extremsten Anwendungsgebieten sein können. Leuchten für den Außenraum sind bei uns mittlerweile wartungsfreie Produkte, die Jahre, sogar Jahrzehnte in Wind und Wetter funktionieren müssen. Korrosionsfreiheit und Gewicht sind von zentraler Bedeutung. So bestimmt das Gewicht die Dimensionierung der Tragstrukturen und ist für uns beim weltweiten Versand auch CO_2-relevant.

Der Scheinwerfer ist ja – wie die meisten Produkte von Erco – formal sehr schön reduziert. Wie wird da dennoch eine Markenidentität erkennbar?
Wenn das Produkt als »formal reduziert« und vielleicht auch noch als »schön« empfunden wird, sind wir im Designteam schon ungeheuer glücklich, denn das gehört zum Kern unseres Corporate Designs. Wir möchten den Kund:innen vorrangig Lichtlösungen liefern. Sind die Produkte später in der Anwendung auch sichtbar, so sollen sie sich ein- oder unterordnen und der Architektur nicht die Show stehlen. Am Ende kommt es aber immer darauf an, dass die Leuchte auch eine gute Figur macht.

→ **Die Außenbeleuchtung ist umstritten – Stichwort Lichtverschmutzung. Warum entwickeln Sie dennoch einen neuen Scheinwerfer?**
Was die ökologischen Aspekte betrifft, stehen heute viele Nutzungen, für die wir Produkte entwickeln, auf dem Prüfstand. In diesem Rahmen suchen wir bessere Lösungen, ohne dabei Sicherheitsaspekte oder heutige Lebensqualität zu vernachlässigen. Im öffentlichen Raum existieren Normen für Beleuchtungsstärken, um sich gefahrlos zu bewegen. Zudem leben urbane Zentren eben auch von Orientierung durch herausgehobene Gebäude oder Plätze bei Nacht. Die Herausforderung besteht heute in der Effizienz, Effektivität und smarten Ansteuerung neuer Beleuchtungskonzepte. Licht soll nur da sein, wo es hingehört und nur dann, wenn man es wirklich benötigt.

Wie eng waren bei der Entwicklung Technik und Design verzahnt?
Beim Design technischer Produkte gilt es immer, die innovativen Möglichkeiten der Technik mit den Erwartungen und Bedürfnissen der Nutzer:innen in Übereinstimmung zu bringen. Der Designprozess ist hier von Beginn an ein Suchen nach der Relevanz und Angemessenheit des Produkts, das entwickelt wird. In einem guten Entwicklungsteam hat jede:r Mitarbeitende diese Sicht verinnerlicht und es bedarf keiner Dominanz von irgendeiner Seite mehr. Wenn Designer:innen in jeden Entwicklungsschritt integriert sind, entsteht auch innerhalb des Teams ein größeres Verständnis darüber, warum die Nutzer:innen und die ökologischen Fragen heute im Mittelpunkt stehen.

Das Familienunternehmen ERCO mit Sitz in Lüdenscheid ist in rund 55 Ländern weltweit vertreten. 1934 gegründet, konnte ERCO in den 1960er-Jahren das neue Feld der Architekturbeleuchtung in Europa etablieren, knapp 50 Jahre später ist das Unternehmen der erste klassische Leuchtenhersteller mit einem komplett auf LED-Technologie basierenden Produktprogramm.

Henk Kosche ist seit 2004 Group Manager Design.

www.erco.com

→ **Light pollution is a serious issue – which is why outdoor lighting is controversial. So why develop a new projector?**
As far as the ecological aspects are concerned, a lot of the uses we develop products for are under scrutiny. Given that context, we're looking for better solutions without neglecting safety aspects or compromising the quality of life people are used to. In public spaces there are norms for illuminance levels so that people can move around safely. And on top of that, urban centres depend on buildings or squares being accentuated at night because they provide orientation. Today the challenge lies in ensuring the efficiency, effectiveness and smart control of new lighting concepts. Light should only be present where it belongs and only when it's really needed.

How closely did engineering and design collaborate on the development?
When it comes to designing technical products, the aim is always to match the innovative technical possibilities with the expectations and needs of the users. Right from the start, the design process is a question of analysing the relevance and appropriateness of the product that's being developed. In a good development team, every single member has internalised that perspective and there's no need for any one side to dominate the other. When designers are integrated into every stage of the development, it also leads to greater understanding within the team itself as to why the focus nowadays is on users and ecological aspects.

You use plastic for the housing – why not cast metal?
Plastics are top-quality materials that can be excellent problem solvers for the most extreme application areas. Our outdoor luminaires are meanwhile maintenance-free products that have to function in all weathers for years, even decades. It's crucial for them to be corrosion-free and lightweight. The weight determines the dimensions of the facade mountings, for instance, and it's relevant in terms of our carbon footprint too because we ship our products all over the world.

The projector – like most of Erco's products – has a very reductive and attractive aesthetic. But how do you reconcile that with a recognisable brand identity?
If the product is perceived as »reductive« and perhaps also »attractive«, we in the design team are already delighted because that's part of the essence of our corporate design. First and foremost, we want to provide our customers with lighting solutions. And if the products are visible when in use, they should blend in or stay in the background rather than trying to steal the show from the architecture. But at the end of the day, it's important for the luminaire to look good as well – that goes without saying.

ERCO is a family business based in Lüdenscheid and represented in approx. 55 countries worldwide. Founded in 1934, it was ERCO that established the new field of architectural lighting in Europe in the 1960s. Today, not quite 50 years later, the company is the first classic luminaire manufacturer with a product portfolio based entirely on LED technology.

Henk Kosche has been Group Manager Design since 2004.

www.erco.com

| SPECIAL MENTION | REFLEX² FLOOR
→ SEITE/PAGE
107 | STEHLEUCHTE
FLOOR LAMP |

SPECIAL MENTION

REFLEX² FLOOR
STEHLEUCHTE / FLOOR LAMP

JURY STATEMENT

Die formal sehr reduzierte Leuchte produziert eine sehr angenehme Lichtstimmung. Im Betrieb verleiht sie dem Licht eine fassbare Quelle, bleibt aber unaufdringlich und fügt sich in nahezu jeden Kontext ein.

A luminaire with an extremely understated design that produces a very pleasant lighting mood. When in use, it serves as an identifiable source of light yet remains unobtrusive and will blend in with virtually any context.

HERSTELLER/MANUFACTURER
Serien Raumleuchten GmbH
Rodgau

DESIGN
Inhouse
Jean Marc da Costa

VERTRIEB/DISTRIBUTOR
Serien Raumleuchten GmbH
Rodgau

Die Ästhetik sachlich und klar, die Form ein Statement: Auf das Äußerste reduzierte Aluminiumprofile, die lediglich die Konturen eines langgestreckten Quaders nachzeichnen, bilden das Rahmengerüst der Stehleuchte. In die Rahmenstruktur eingebettete LED-Platinen geben das Licht über die Decke in den Raum ab und sorgen für eine rundum indirekte, blendfreie Beleuchtung. Die gewünschte Helligkeit lässt sich über einen Fußtaster im Sockel regeln. Die Leuchte steht in zwei Größen und den Farben Schwarz und Weiß zur Verfügung, das größere Modell wartet zudem mit einer variablen Farbtemperatursteuerung auf.

The aesthetic is objective and clear, the form a statement: the floor lamp's frame structure is constructed out of aluminium profiles that have been reduced to the absolute minimum and merely trace the contours of an elongated rectangular cuboid. LED boards embedded in the frame cast their light into the room via the ceiling to provide indirect, glare-free lighting all round. A foot switch in the base is used to regulate brightness. The lamp is available in two sizes and with a black or white frame; in addition, the colour temperature of the larger model is variable.

CONSUMERELECTRONIC, ENTERTAINMENT
CONSUMER ELECTRONICS, ENTERTAINMENT

1 **SPECIAL MENTION:**
COUCOU
Coucou GmbH
Essen
und/and
Image Construction GmbH
Erkelenz

Während die klassischen Gerätschaften allmählich durch digitale Devices und Services verdrängt werden, tauchen am Firmament ganz neue Anforderungen auf, getragen durch Vernetzung, Heimarbeit oder intelligente Systeme, für die keine gestalterischen Archetypen existieren. Hier setzt das Design komplett neu an.

While classic appliances are gradually being superseded by digital devices and services, totally new needs are appearing on the horizon, driven by networking, homeworking or intelligent systems for which no archetypes exist. In cases such as these, design has to start from scratch.

SPECIAL MENTION | COUCOU → SEITE/PAGE 111 | VIDEOKABINE / VIDEO CABIN

SPECIAL MENTION — COUCOU — VIDEOKABINE / VIDEO CABIN

JURY STATEMENT

Die Kabine bietet mit ihrer technischen Ausstattung die besten Voraussetzungen für ergonomische und imagefördernde Video-Calls. Eine insgesamt sehr spannende Konzeption, die auch Orientierung in Open Spaces sowie Rückzugsort für konzentriertes, hybrides Arbeiten sein kann.

The cabin and its technical features provide ideal conditions for ergonomic and image-friendly video calls. All in all a very intriguing concept that can provide orientation in open spaces as well as a refuge for focused, hybrid work.

HERSTELLER/MANUFACTURER
Coucou GmbH
Essen
und/and
Image Construction GmbH
Erkelenz

DESIGN
Middelhauvedesign
Essen

VERTRIEB/DISTRIBUTOR
Coucou GmbH
Essen

In Video-Meetings schalten sich die Teilnehmer:innen oft aus improvisierten Umgebungen zu; das Einkopieren eines virtuellen Hintergrunds schafft meist keine professionelle Anmutung. Aus diesem Grund hat das Essener Start-Up eine optimierte Umgebung für qualitativ bessere Video-Calls entwickelt. Die kompakte Kabine trennt akustisch von der Umgebung, bietet aber durch Glasflächen Sichtbezüge nach draußen. Im Inneren wartet eine technische Grundausstattung in Form von Kamera, Monitor, Lautsprecher, gesichtsfreundlicher Ausleuchtung sowie höhenverstellbarem Arbeitsplatz auf die temporären Nutzer:innen, die ihr persönliches Notebook oder Tablet über Schnittstellen anbinden. Der Hintergrund ist optisch ruhig gehalten, das Äußere lässt sich farblich individualisieren. Die Belüftung der Kabine erfolgt automatisch und lässt sich mit einer Desinfektionsstufe ergänzen.

Participants often join video meetings from improvised settings; copying in a virtual background is generally not enough to create a professional impression. That was what prompted the Essen-based startup to develop an optimised environment for better-quality video calls. Inside the compact cabin, the occupant is isolated from ambient noise but can maintain visual contact with their surroundings thanks to the large areas of glass. The interior provides basic technical equipment in the form of a camera, monitor, speaker, face-friendly lighting and a height-adjustable workstation with ports for the temporary user's personal laptop or tablet. The backdrop is sedate, whereas the colour of the exterior is customisable. The automatic ventilation comes with an optional disinfection stage.

ANDREAS HESS **WHITE ID, SCHORNDORF**

»Gestaltung muss in Bezug auf Nachhaltigkeit künftig noch bessere Produktlösungen liefern, die durch Funktion, Nutzerbezug, Langlebigkeit und ihr hohes Maß an Emotionalität den Konsumenten selbstverständlich überzeugen.«

»In terms of sustainability, design will have to deliver even better product solutions in future – products that convince the consumer self-evidently through their function, connection with the user, longevity and high degree of emotionality.«

Andreas Hess ist Diplom-Produktdesigner und seit 2002 Geschäftsführer der Designagentur White ID GmbH & Co. KG. Das Büro wurde bereits mehrfach mit dem Focus Open ausgezeichnet, unter anderem für die ganzheitliche Konzeption von Kindersitzen für Fahrzeuge. Hess war unter anderem 14 Jahre Leiter für Design und Brand-Management beim Kindersitz-Hersteller Concord. Seit 2021 engagiert er sich als Gastprofessor für Produktgestaltung an der HfG Schwäbisch Gmünd.

Andreas Hess trained as a product designer and has been a managing partner at design agency White ID GmbH & Co. KG since 2002. The firm has won a number of Focus Open awards over the years, including for its holistic concept for child car seats. Among other things, Hess spent 14 years as head of design and brand management for child car seat manufacturer Concord. Since 2021, he has been a visiting professor in product design at the University of Design (HfG) Schwäbisch Gmünd.

www.white-id.com

www.white-id.com

1 → SEITE/PAGE
116–121

2 → SEITE/PAGE
122, 126

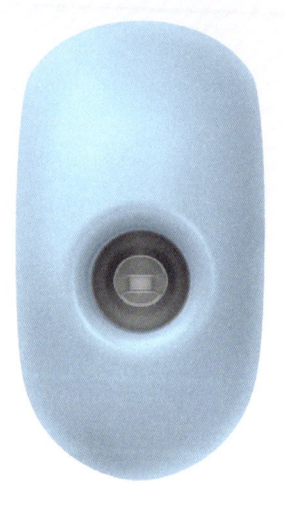

3 → SEITE/PAGE
123, 127

4 → SEITE/PAGE
124, 128

5 → SEITE/PAGE
125, 129

FREIZEIT, SPORT, SPIELEN
LEISURE, SPORTS, PLAY

GOLD:
1 **TRANSALP PRO**
Fischer Sports GmbH
Ried im Innkreis
Österreich/Austria

SILVER:
2 **STOFFBOOT**
Stoffboot
Leipzig

3 **SUGAR RUSH**
Triple A Marketing GmbH
Bielefeld

4 **FANOM**
Speedfab GmbH
Schorndorf

SPECIAL MENTION:
5 **M99 PRO2**
Supernova Design GmbH
Gundelfingen

Einst Leerraum für Entspannung, Rekreation und absichtsloses Sein, ist die Freizeit längst prall gefüllt mit Aktivitätsangeboten und entsprechenden Produkten für unterschiedlichste Neigungen, Altersgruppen und Erlebnisversprechen. Schön, wenn es da Dinge gibt, die keinen unmittelbaren Zweck erfüllen wollen – oder das Naturerlebnis intensivieren.

Whereas free time was once a vacuum waiting to be filled with relaxation, recreation and delectable idleness, today it is crammed full with a vast spectrum of activity options and the corresponding products for diverse inclinations, age groups and experiences. Every now and again, it's nice when there are things that serve no direct purpose – or intensify the way we experience nature.

GOLD | TRANSALP PRO | TOURENSKISCHUH / SKI TOURING BOOT

TRANSA PRO TOURENSKI SCHUH

FREIZEIT, SPORT, SPIELEN
LEISURE, SPORTS, PLAY

FOCUS GOLD

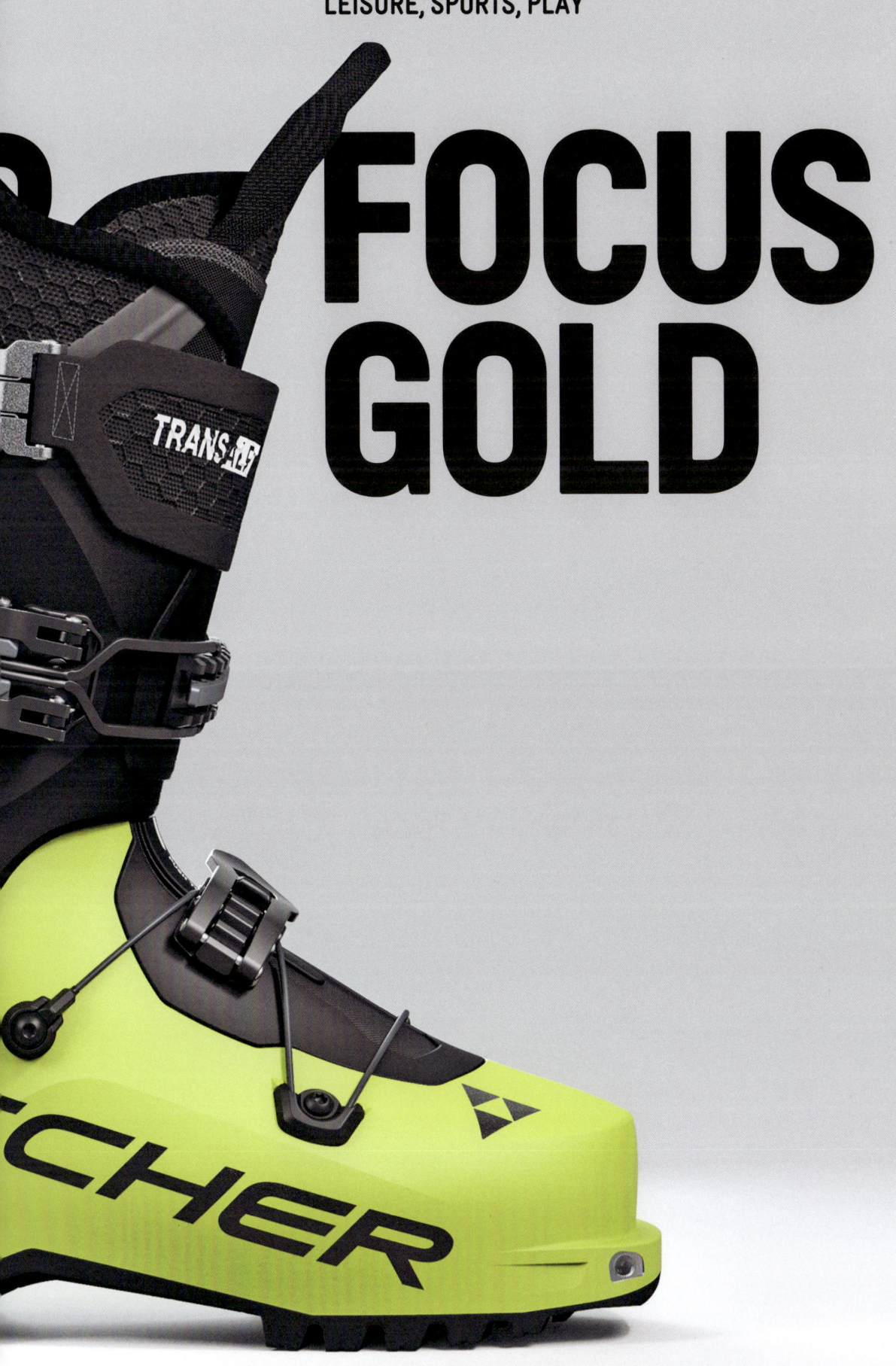

GOLD — TRANSALP PRO — TOURENSKISCHUH / SKI TOURING BOOT

JURY STATEMENT

Die Gestaltung überzeugt in mehrerer Hinsicht. Der Schuh mit seiner schmalen Silhouette präsentiert sich sehr dynamisch, ohne dabei überzogene Stylingelemente zu nutzen. Alle Formübergänge sind sehr spannungsvoll und sensibel modelliert, die Farbgebung erzeugt Spannung. Sehr gut gelöst wurden überdies die leicht bedienbare Schnallenmechanik und die Schaftfreigabe.

The design is compelling in several respects. Thanks to its slender silhouette, the boot makes an extremely dynamic impression without resorting to exaggerated styling elements. All the transitions between the various forms are suspenseful and sensitive, the use of colour generates a sense of excitement. In addition, the solutions found for the easy-to-use buckle mechanism and cuff release are very compelling.

HERSTELLER/MANUFACTURER
Fischer Sports GmbH
Ried im Innkreis
Österreich/Austria

DESIGN
Formquadrat GmbH
Linz
Österreich/Austria

VERTRIEB/DISTRIBUTOR
Fischer Sports GmbH
Ried im Innkreis
Österreich/Austria

Skitouren haben in den letzten Jahren enorm an Popularität gewonnen. Da Skitourengeher:innen sowohl bergauf als auch bergab unterwegs sind, muss sich die Ausrüstung an die jeweilige Situation anpassen. So lässt sich der Schaft des Transalp-Tourenschuhs für den Aufstieg mit einem Handgriff lösen, um die Rotation des Sprunggelenks um 80 Grad zu ermöglichen. Abwärts wiederum bleibt der Schaft fixiert, die Beinkraft wird dann optimal auf den Ski übertragen. Die Neigung des Schafts lässt sich – je nach sportlichen Ambitionen – zwischen 13 und 17 Grad wählen. Insgesamt wiegt der Schuh nur 1,28 Kilogramm, was vor allem der steifen und dünnen Schale zu verdanken ist. Das Design unterstreicht diesen Reduktionsansatz, etwa mittels der filigranen, einhändig bedienbaren Schnallen. Der Schuh setzt die Designlinie des Vorjahresmodells fort und ist in vier Varianten erhältlich – darunter auch eine für schmale Füße.

Ski touring has seen a huge rise in popularity in recent years. And because it involves skiing both uphill and downhill, the equipment has to adapt to the situation at any given time. For the ascent, for instance, the cuff of the Transalp touring boot can be released in one easy move, permitting 80-degree rotation of the ankle articulation. On the descent, however, the cuff remains fixed in place, resulting in the optimal transmission of force from leg to ski. Depending on how ambitious the user is, the forward lean of the cuff can be adjusted between 13 and 17 degrees. The entire boot weighs just 1.28 kilograms, mainly thanks to the thin yet stiff shell. The design underscores this reductivist approach with features such as the filigree one-handed buckles. The boot is an evolution of the previous year's model and available in four versions – including one for narrow feet.

STEFAN DEGN   **GESELLSCHAFTER, FORMQUADRAT GMBH**

»Wenn man ein Produkt selbst nutzt, hat man ein viel tieferes Verständnis für Funktionen und Pain Points.«

»When you use a product yourself, you have a much deeper understanding of the functions and pain points.«

STEFAN DEGN — PARTNER, FORMQUADRAT GMBH

→ **Wie herausfordernd ist das Design eines Skischuhs?**

Ein Tourenskischuh ist enormen Kräften ausgesetzt und muss entsprechend stabil sein. Gleichzeitig soll der Schuh leicht und bequem sein sowie ergonomische Anforderungen erfüllen. Auch das Verschlusssystem soll einfach zu bedienen sein, dennoch muss es gut schließen und den Schuh zusätzlich stabilisieren. Diese Faktoren schaffen ein Spannungsfeld, in dem wir uns als Designer:innen bewegen. Die Herausforderung ist, allen Voraussetzungen in einer maximal ansprechenden Form gerecht zu werden.

Muss man selbst Tourengeher sein, um zu wissen, worauf es ankommt?

Wenn man ein Produkt selbst nutzt, hat man ein viel tieferes Verständnis für Funktionen, Pain Points, Verstaubarkeit, Nutzungsrituale. Das eigene Erleben schafft einen ganz anderen emotionalen Bezug während der Entwicklung des Produkts. Wir arbeiten von Linz aus und haben die Alpen vor der Haustüre – dementsprechend haben wir begeisterte Tourengeher:innen im Team, die Produkte wie den Transalp nicht nur gestalten, sondern auch regelmäßig selbst nutzen.

Wie interdisziplinär ist die Entwicklung eines solchen Sportprodukts?

Wintersportler:innen, das Produktmanagement und die Produktentwicklung, aber auch Fertigung, Vertrieb und wir als Gestaltende bringen individuelle Perspektiven ein – und zwar zu Beginn des Prozesses. Verschiedene Blickwinkel führen zu iterativen Schleifen und Optimierungen im Gestaltungsprozess, an dessen Ende ein noch besseres Produkt steht.

Welche Rolle spielt die Werkstoffentwicklung für das Design?

Vor allem im Sportbereich passiert hier jede Menge, es kommen stetig neue Werkstoffe und Materialien auf den Markt. Dabei spielen auch die Themen Nachhaltigkeit und Wiederverwertbarkeit eine große Rolle. Bei der Entwicklung neuer Produkte ist es spannend, ob und welche neuen Materialien zum Einsatz kommen könnten. Bei Fischer wird beispielsweise der Kunststoff Pebax auf Basis von Rizinusöl verwendet. Das hat natürlich auch für uns Folgen. Je nach Werkstoff kann das gestalterische Einschränkungen bedeuten, genauso aber neue Möglichkeiten eröffnen.

Und wie sehr mussten Sie dabei Branding-Vorgaben integrieren?

Der Transalp ist nicht der erste Skischuh von Fischer aus unserer Feder, daher konnten wir die von uns initiierte Gestaltungsrichtung weiterentwickeln. Fischer-Produkte sollen auch ohne Logo, nur durch Form und Linienführung, eindeutig als solche erkennbar sein. Dennoch kommen Vorgaben aus der Produktgrafik. Der Fischer-Schriftzug ist ein integrativer Teil des Gesamtkonzepts. Das wird schon bei der Gestaltung der Flächen und Kanten mitgedacht.

Gegründet von Stefan Degn und Mario Zeppetzauer, steht Formquadrat mit Standorten in Linz und Gmunden seit über 20 Jahren für die Gestaltung technischer Produkte, die vielfach ausgezeichnet und in den Märkten erfolgreich sind.

www.formquadrat.com

→ **How challenging is it to design a ski boot?**

A ski touring boot is subject to enormous forces so it has to have the necessary stability. At the same time the boot needs to be light and comfortable and meet the ergonomic requirements. The fastening system also has to be easy to use, while nevertheless closing securely and additionally stabilising the boot. Taken together, these factors define the limits within which we designers are free to move. The challenge is to find a solution that meets all the requirements in the most appealing form possible.

Do you have to be a touring skier yourself to know what's really important?

When you use a product yourself, you have a much deeper understanding of the functions and pain points. Your own experience results in a totally different emotional relationship to the product's development. We're based in Linz so the Alps are on our doorstep – and there are some keen touring skiers in the team who don't just design products like Transalp, they use them themselves on a regular basis.

How interdisciplinary is the development of a sports product like this?

Winter sports enthusiasts, product management and product development, but also the manufacturer, sales and we designers all play a role and contribute our individual perspectives – that happens right at the start of the process. All those different points of view lead to iterative loops and optimisations in the design process, which culminates in an even better product.

What role does materials development play for the design?

There's a lot happening in that respect, especially in the sports sector, new materials are coming onto the market all the time. Sustainability and recyclability play a big role as well. When you're developing new products it's exciting to see if you could use new materials for them, and if so which ones. At Fischer, for instance, they use a plastic called Pebax, which is based on castor oil. That obviously has consequences for us too. The choice of material can result in limitations for the design, but it can just as easily open up new possibilities as well.

And to what extent did you have to integrate branding guidelines?

The Transalp isn't the first ski boot we've done for Fischer, so we were able to evolve the design direction that we'd already initiated. Fischer products should be clearly recognisable as such even without a logo, just by their shape and lines. Even so, there are still guidelines in terms of product graphics. The Fischer logotype is an integral part of the overall concept and is already factored in to the equation when we're designing the surfaces and edges.

Founded by Stefan Degn and Mario Zeppetzauer and with locations in Linz and Gmunden, Formquadrat has been designing award-winning and commercially successful technical products for more than 20 years.

www.formquadrat.com

SILVER   STOFFBOOT   KANU
         → SEITE/PAGE   CANOE
         126

SILVER	SUGAR RUSH	SEXTOY	
	→ SEITE/PAGE	SEX TOY	
	127		

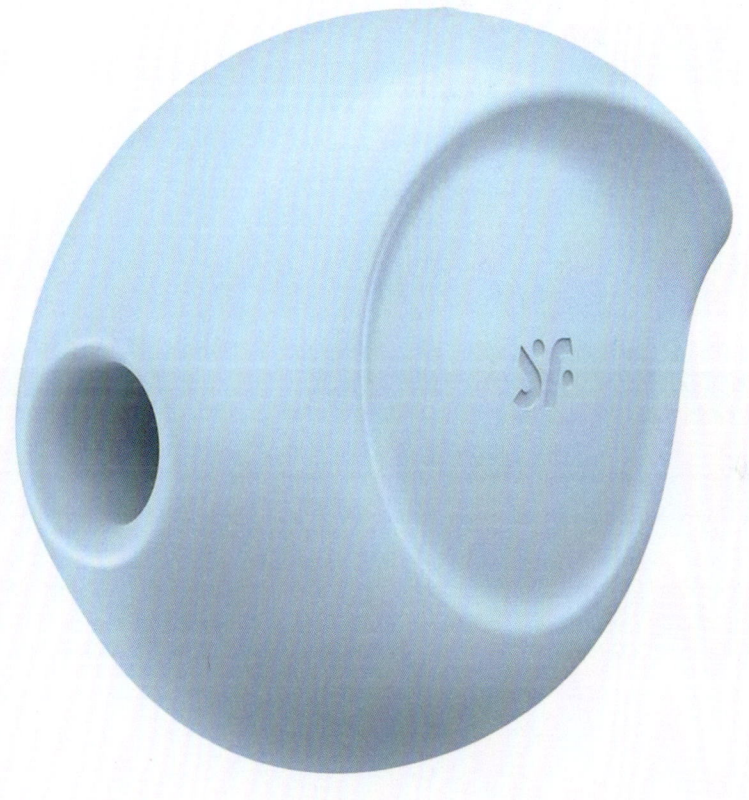

SILVER FANOM
→ SEITE/PAGE 128

KÖRPERPROTEKTOREN
BODY PROTECTORS

SILVER

STOFFBOOT

KANU
CANOE

JURY STATEMENT

Ein ehrliches, zurückhaltendes Produkt, das handwerklich sehr gut verarbeitet ist. Das Boot brilliert mit geringem Gewicht und einer harmonischen Anmutung. Obwohl der Rumpf voluminös ist, lässt er sich im Handumdrehen fahrradkompatibel klappen. Das verspricht ein gelungenes Freizeiterlebnis.

An honest, understated design paired with very good craftsmanship. The boat stands out for its low weight and harmonious appearance. Despite its voluminous hull, it can be folded up ready to be towed by a bicycle in just a few simple moves – all of which adds up to the promise of a highly satisfying leisure experience

HERSTELLER/MANUFACTURER
Stoffboot
Leipzig

DESIGN
Inhouse
Ringo Köhler

VERTRIEB/DISTRIBUTOR
Stoffboot
Leipzig

Die Lust am sportiven Draußensein ist ungebrochen. Neben Radfahren boomen Freizeitbetätigungen auf dem Wasser wie SUP, Paddeln oder Flusswandern – immer mit dem Problem verbunden, am Endpunkt der Tour Mensch und Gerät wieder per Auto, Bus oder Bahn zurückbringen zu müssen. Diesem Umstand bereitet das in zwei Hälften teilbare Stoffboot ein Ende. Mit einem Gewicht von nur zehn Kilogramm lässt sich jede Hälfte problemlos tragen, um Ecken manövrieren und fahrradtauglich anhängen. Für das wendige Boot kommt eine Skin-on-Frame-Bauweise zum Einsatz. Der Rumpf besteht aus Sperrholz regional verfügbarer Hölzer, die Bespannung aus schwerem, in Leinölfirnis getränktem Canvasgewebe. Zusammengebaut ist das Boot ganz fix und werkzeuglos: Zwei Steckscharniere und sechs einfache Verschraubungen sorgen für eine sichere Verbindung beider Hälften.

The desire to indulge in outdoor sports remains unabated. In addition to cycling, water-based leisure activities such as SUP, canoeing and recreational kayaking are booming – and always come with the problem of having to get both the equipment and its users back to the car, bus or train once the tour is over. Because it can be divided into two halves, this fabric boat puts an end to that dilemma. Weighing in at just 10 kilograms each, the two halves can be carried, manoeuvred round corners and towed by a bicycle without any problem at all. The agile boat is a skin-on-frame construction. The hull consists of plywood made from regionally sourced wood, the covering is made of heavy canvas soaked in linseed oil varnish. The boat can be put together in next to no time, no tools required: two hinges and six simple screws ensure the two halves are securely connected.

SILVER — SUGAR RUSH — SEXTOY / SEX TOY

JURY STATEMENT

Ein Sextoy, das sich im ersten Moment nicht als solches zu erkennen gibt. Es ist einerseits diskret, andererseits zeigt es eine ganz eigene Ästhetik und liegt gut in der Hand. Gelungen ist auch die Reduktion auf ein Material.

A sex toy that isn't immediately recognisable as such. Although discreet, it nevertheless has a distinctive aesthetic and sits comfortably in the hand. The fact that the design has been reduced to a single material is another commendable aspect.

HERSTELLER/MANUFACTURER
Triple A Marketing GmbH
Bielefeld

DESIGN
Inhouse

VERTRIEB/DISTRIBUTOR
Triple A Marketing GmbH
Bielefeld

Seinerzeit ein wirkliches Novum auf dem Markt der Sextoys, bereichert die berührungslose Stimulation der weiblichen Klitoris per Druckwellentechnologie mittlerweile das Liebesleben zahlreicher Frauen, Männer und Paare. Der in seiner kompakten Form an einen kleinen Fisch erinnernde Satisfyer aus medizinischem Silikon kann sogar noch mehr: Elf verschiedene Druckwellenintensitäten lassen sich mit zwölf Vibrationsprogrammen kombinieren oder auch getrennt ansteuern. Ein integrierter Akku sorgt für umweltfreundliches Wiederaufladen des nahtlos verarbeiteten und wasserdichten Toys.

A genuine novelty on the sex toy market when it was first launched, the use of pressure wave technology for non-contact stimulation of the female clitoris has been enriching the sex lives of numerous women, men and couples ever since. Made of medical-grade silicone and with a compact form reminiscent of a little fish, this satisfyer goes far beyond the basics: thanks to its two motors, the 11 different pressure wave intensities and 12 vibration programmes can either be combined or controlled separately. The seamless and waterproof toy is equipped with an integrated battery for eco-friendly recharging.

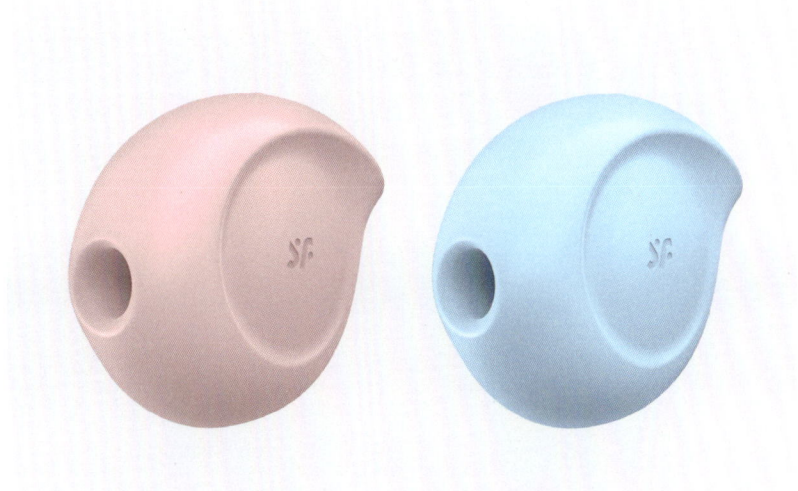

SILVER

FANOM
KÖRPERPROTEKTOREN
BODY PROTECTORS

> **JURY STATEMENT**
>
> Sehr anwendungsfreundliche und sinnvolle Schutzelemente für Sportler:innen. Sie schränken die Beweglichkeit kaum ein und müssen durch ihre interessante formale Umsetzung auch nicht unbedingt versteckt werden.
>
> Very practical and sensible protective elements that barely restrict freedom of movement at all during sporting activities. Thanks to their interesting design, the user won't necessarily want to conceal them either.

HERSTELLER/MANUFACTURER
Speedfab GmbH
Schorndorf

DESIGN
DQBD GmbH
Schorndorf

VERTRIEB/DISTRIBUTOR
Speedfab GmbH
Schorndorf

Nicht nur bei ausgesprochenen Risikosportarten ist der Schutz von Gelenken, des Rückens oder der Brust mit energieabsorbierenden Protektoren sinnvoll. Die Protektoren der Reihe »Fanom« nutzen dafür eine Y-Struktur, deren Elemente stets senkrecht zum Körper stehen und damit die Sturzenergie optimal aufnehmen. Außerdem sorgt die Struktur für gute Ventilation, passt sich den individuellen Physiognomien an und trägt nicht dick auf. Alle geometrisch parametrisierbaren Protektoren bestehen aus recycelbarem TPU, das im Spritzguss verarbeitet wird. Die Protektoren können in Kleidungsstücke eingelegt, aufgenäht oder mit Hilfe von Klettbändern fixiert werden.

Taking part in high-risk sports is by no means the only situation where cushioning the back, chest and joints with energy-absorbing protectors makes a lot of sense. The protectors of the Fanom series are based on a structure made up of Y-shaped elements that are always vertical to the body, thereby ensuring that the impact of the fall is optimally absorbed. In addition, the structure ensures good ventilation, adapts to the user's individual physiognomy and is not bulky. The parametrisable protectors are made of injection-moulded and recyclable TPU and can be inserted into clothes, sewn on or fixed in place with Velcro straps.

SPECIAL MENTION — M99 PRO2 — E-BIKE-SCHEINWERFER / E-BIKE HEADLIGHT

JURY STATEMENT

Sauber gestalteter Frontscheinwerfer für E-Bikes, der durch sein integriertes neuartiges Energiemanagement für mehr Sicherheit sorgt – er verbessert die Sichtbarkeit der Radfahrenden im Verkehr und sorgt bei nächtlichen, schnellen Pendlerfahrten für beste Ausleuchtung des Terrains.

A neatly designed front light for e-bikes that uses innovative integrated energy management to provide greater safety – it improves cyclists' visibility in traffic and ensures optimal illumination of the terrain during high-speed nocturnal commutes.

HERSTELLER/MANUFACTURER
Supernova Design GmbH
Gundelfingen

DESIGN
Inhouse

VERTRIEB/DISTRIBUTOR
Supernova Design GmbH
Gundelfingen

Der Scheinwerfer nimmt die Formensprache der M99-Reihe auf und erweitert die visuelle Präsenz durch eine ringförmige Tagfahrlicht-Signatur, die per Sensor automatisch aktiviert wird – und über eine offizielle Zulassung verfügt. Über zwei kleine, geschützte Buchsen an der Rückseite lassen sich nachträglich Zusatzgeräte anschließen, die in das softwaregesteuerte Energiemanagement einbezogen werden. Dieses System, die eigentliche Innovation des Scheinwerfers, sorgt dafür, dass die unterschiedlichen elektrischen Komponenten eines E-Bikes jeweils nur so viel Strom erhalten, wie momentan nötig ist. Während konventionelle Systeme für die einzelnen Verbraucher feste Energiekontingente reservieren, wird die Energieversorgung nun für ein starkes Abblend- und Fernlicht priorisiert. Kurzfristige Verbrauchsspitzen puffert ein zusätzlicher Speicher im Scheinwerfer ab.

The headlight picks up on the design language of the M99 series and amplifies its visual presence with a ring-shaped daytime running light signature that is automatically activated via a sensor – and has been officially approved. Two small, protected ports on the back allow the subsequent incorporation of additional features that are then included in the software-based energy management system. This system, which constitutes the headlight's actual innovation, ensures that the e-bike's various electrical components are only supplied with as much power as is necessary at any given moment. Whereas conventional systems reserve fixed amounts of energy for the individual loads, in this case the energy supply is prioritised to ensure powerful low- and high-beam light. An additional reserve in the headlight serves as a buffer for short-term peaks in energy consumption.

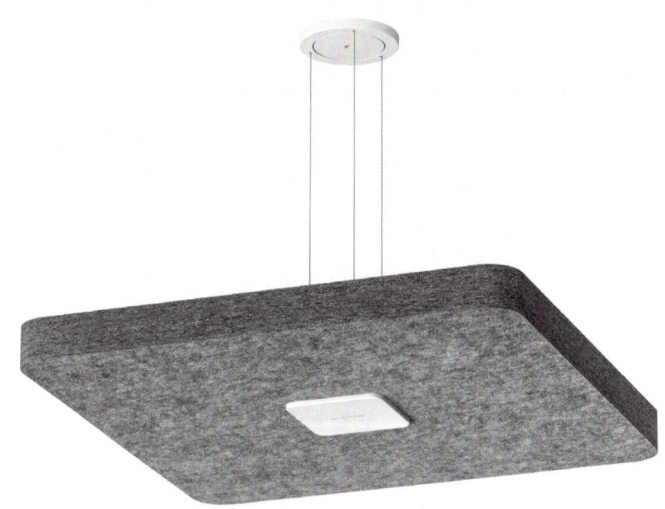

1 → SEITE/PAGE
   134, 136

2 → SEITE/PAGE
   135, 137

GEBÄUDETECHNIK
BUILDING TECHNOLOGY

SILVER:
1 **XL-SERIE**
Primo GmbH
Aschau am Inn

SPECIAL MENTION:
2 **DISC'N DOTS**
Nimbus Group GmbH
Stuttgart

Design unterstützt Architektur in ihrem Bestreben, ästhetische und zugleich funktionale, effiziente sowie zukunftstaugliche Gebäude zu realisieren. Dazu gehören die weiten Welten der Haustechnik und Gebäudesteuerung, aber auch das Universum der Ausstattung oder der Produkte, die lediglich in der Bauphase in Erscheinung treten.

Design supports architecture in its endeavours to create aesthetic buildings that are nevertheless functional, efficient and futureproof. That applies not just to the broad fields of domestic technology and building automation, but also to the world of equipment and products that are only in evidence during the construction phase.

| SILVER | XL-SERIE | EINBAUGEHÄUSE |
|||INSTALLATION HOUSINGS|

→ SEITE/PAGE
136

| SPECIAL MENTION | DISC'N DOTS → SEITE/PAGE 137 | AKUSTIKELEMENT ACOUSTIC ELEMENT |

SILVER XL-SERIE EINBAUGEHÄUSE / INSTALLATION HOUSINGS

> **JURY STATEMENT**
>
> Ein absolut nutzerzentriertes Produkt, das die Anforderungen einer agilen Elektroinstallation bestens erfüllt – und die Rohrdurchführungen farblich sofort erkennbar macht. Das Orange leitet sich aus dem Corporate Design ab und stärkt die Sichtbarkeit der Marke im rauen Alltag auf der Baustelle.
>
> A thoroughly user-centric product that more than meets the requirements for the agile installation of electrical wiring – and uses colour to make the conduit entry points immediately recognisable. The orange is an element of the corporate design and strengthens the brand's visibility in the tough everyday context of the construction site.

HERSTELLER/MANUFACTURER
Primo GmbH
Aschau am Inn

DESIGN
Andreas Schulze/Industrial Design
Limburg an der Lahn

VERTRIEB/DISTRIBUTOR
Primo GmbH
Aschau am Inn

Die Gehäusefamilie wurde entwickelt, um die Elektroinstallation in Betondecken oder -wänden zu vereinfachen. Die Gehäuse mit andockbarem Trafotunnel werden auf die Schalungen gelegt und mit Nägeln, Kleber oder per Magneteinsatz befestigt. Anschließend erfolgt das Verlegen der Leerrohre für die spätere Kabelzuführung. Die Rohre werden durch die vorgesehenen, mit TPE-Membranen verschlossenen Öffnungen geschoben, innen automatisch per Rohrstopp fixiert und erhalten so eine sichere Zugentlastung. Betondicht und trittfest eignen sich die Gehäuse für die Ortbeton- wie auch Fertigteil-Bauweise. Ist die Rohdecke fertig, entfernt man die auf der Deckenunterseite plan liegende Frontplatte und kann so von unten Kabel, Leuchten, Lautsprecher oder andere Einbaugeräte installieren.

Das Design nutzt gespannte Flächen und die Hausfarbe, um Akzente zu setzen und die Rohrdurchführungen zu markieren.

The series of housings was developed with the goal of simplifying the installation of electrical wiring in concrete ceilings or walls. The housings, which can be equipped with a dock-on universal tunnel, are placed on the formwork and attached with nails, adhesive or a magnetic insert. In the next step, the conduits for the cable routing are laid. The conduits are pushed through the openings, which are sealed with TPE membranes, and automatically fixed in place inside the housing by a grip that ensures reliable strain relief. Concrete-tight and walkable, the housings are suitable for both cast-in-place and precast concrete. Once the ceiling slab is finished, the front plate on the underside of the ceiling is removed ready for the installation of cables, lamps, speakers or other recessed devices.

The design uses curved surfaces and the corporate colour to create interesting accents and indicate the entry points for the conduits.

SPECIAL MENTION — DISC'N DOTS

AKUSTIKELEMENT / ACOUSTIC ELEMENT

JURY STATEMENT

Die reduzierten Formen fügen sich bestens in die allgemeine Designsprache des Herstellers ein, Akustikelemente und Leuchten basieren auf der gleichen minimalistischen Grundidee. Sehr gut gelöst ist die Befestigung, die nicht mehr versteckt wird, sondern als visueller Akzent eine wichtige Rolle im Gesamtkonzept spielt.

The understated forms blend in perfectly with the manufacturer's overarching design language: both the acoustic elements and luminaires are based on the same minimalist idea. A very good solution has been found for the mounting: rather than being concealed, it now serves as a visual highlight that plays an important role in the overall concept.

HERSTELLER/MANUFACTURER
Nimbus Group GmbH
Stuttgart

DESIGN
ID Aid GmbH
Stuttgart

VERTRIEB/DISTRIBUTOR
Nimbus Group GmbH
Stuttgart

In vier Formen und Größen erhältlich, lassen sich die Akustikelemente frei im Raum kombinieren, entweder direkt an die Decke montiert oder dekorativ abgependelt. In jedem Fall optimieren die Schallabsorber der Klasse A die Akustik in Räumen mit harten Oberflächen spürbar. Die in Grau oder Weiß erhältlichen, verschnittarm produzierten Module bestehen aus Vlies mit einem 30-prozentigen Recyclingfaseranteil und entsprechen dem Oeko-Tex Standard 100. Die Montage erfolgt über nur einen Befestigungspunkt mittels der »Dots« genannten Elemente von der Sichtseite des Moduls her. Zugleich beleben die Dots mit ihren warmen Farben die monochromen Platten – und sind auch für individuell geformte Akustikelemente nutzbar.

Available in four shapes and sizes, the acoustic elements can be used anywhere in the room and combined as required, either mounted directly on the ceiling or suspended to create a decorative effect. Either way, the Class A sound absorbers noticeably improve the acoustics in rooms with hard surfaces. Produced in such a way that offcuts are minimised, the grey or white modules are made of a fleece that contains 30% recycled fibres and therefore comply with the Oeko-Tex Standard 100. The modules are installed from the visible side, using a single-point mounting inserted via the Dots. At the same time, the warm colours of the Dots add a lively touch to the monochrome panels – and can also be used in conjunction with individually shaped acoustic elements.

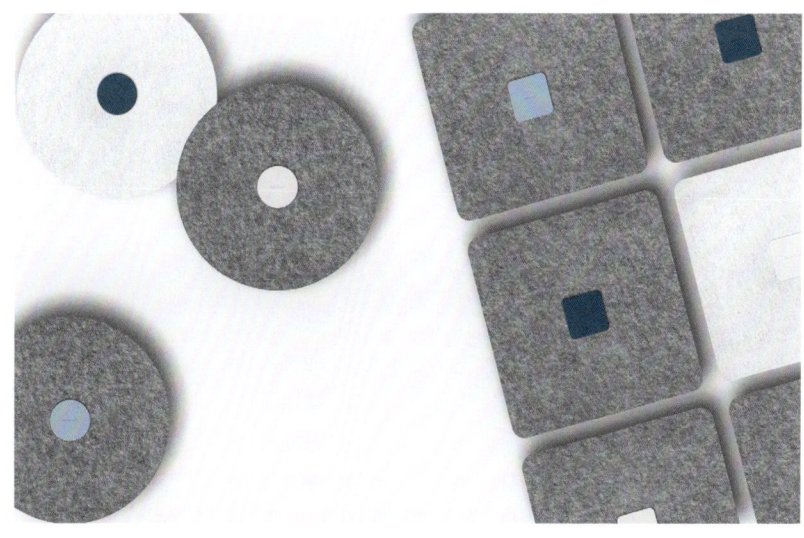

MARC-GREGOR WEIDT EINMALEINS GMBH,
BURGRIEDEN

»Da der ökologische Fußabdruck eines Produktes zum Großteil bereits in der Entwicklungsphase festgelegt wird, ist es essenziell, das Design so früh wie möglich einzubinden.«

»Since a product's ecological footprint is largely determined in the development phase, it's essential to involve design as early on as possible.«

Marc-Gregor Weidt hat ursprünglich das Handwerk des Täschners erlernt, dann an der HfG Schwäbisch Gmünd Produktgestaltung studiert. 1995 gründete er das Büro für Gestaltung Einmaleins, das seit 2016 als Einmaleins GmbH firmiert und dessen geschäftsführender Gesellschafter Weidt ist. Das 16-köpfige Designteam ist interdisziplinär aufgestellt und widmet sich dem Produktdesign, Grafikdesign, UI-Design und Visualisierungskonzepten.

Marc-Gregor Weidt originally learned the craft of bag making before studying product design at the University of Design (HfG) Schwäbisch Gmünd. In 1995 he founded the design firm Büro für Gestaltung Einmaleins, which has operated under the name Einmaleins GmbH since 2016 and is headed by Weidt as managing partner. The 16-strong design team spans various disciplines and focuses on product design, graphic design, UI design and visualisation concepts.

www.einmaleins.net

www.einmaleins.net

1 → SEITE/PAGE
142, 146

2 → SEITE/PAGE
143, 147

3 → SEITE/PAGE
144, 148

4 → SEITE/PAGE
145, 149

5 → SEITE/PAGE
152–159

PUBLIC DESIGN, URBAN DESIGN
PUBLIC DESIGN, URBAN DESIGN

SILVER:
1. **MAGISCHES LAND**
   Archäologisches Landesmuseum
   Baden-Württemberg
   Konstanz

2. **15:14**
   Haus der Geschichte
   Baden-Württemberg
   Stuttgart

3. **ANTHROPOZÄN**
   Staatliches Museum
   für Naturkunde
   Stuttgart

SPECIAL MENTION:
4. **CLIC UP**
   Burkhardt Leitner
   Modular Spaces GmbH
   Stuttgart

META:
5. **URBANHARBOR**
   maxmaier urbandevelopment
   Ludwigsburg

Gestaltung für die Öffentlichkeit ist immer auch von der Inszenierung geprägt – besonders Ausstellungskonzepte mit ihren multimedialen Präsentationen von Exponaten und Geschichten bedienen sich dieses Prinzips. Aber auch der öffentliche Raum gewinnt durch gestalterisch durchdachte Systeme an Attraktivität, Transparenz und Vielfalt.

Design for the public is always influenced by the need to stage things to some extent – exhibition concepts are particularly likely to adopt this principle in the form of multimedia exhibits and narratives. But well-designed systems add to the attractiveness, transparency and diversity of the public space too.

SILVER

MAGISCHES LAND
AUSSTELLUNGSKONZEPTION
EXHIBITION CONCEPT

JURY STATEMENT

Dieser Ausflug in die Frühzeit ist besonders faszinierend, weil die Geschichte atmosphärisch erlebbar wird. Die Aufarbeitung des zunächst sehr trockenen Themas ist sehr gelungen, die Einbindung multimedialer Elemente ebenso gut gelöst wie der gezielte Einsatz von Licht.

This excursion into ancient times is particularly fascinating because it uses atmosphere as a medium for bringing history to life. A very successful presentation of what initially seems like a very dry topic: the integration of multimedia elements is just as compelling as the targeted use of light.

AUFTRAGGEBER/CLIENT
Archäologisches Landesmuseum Baden-Württemberg
Konstanz/Constance

DESIGN
Inhouse
Simon Neßler

FOTOS/PHOTOS
Daniel Strauch (S/P 142)
Ben Wiesenfarth (S/P 146)

Das Land Baden-Württemberg fördert mit der sogenannten Keltenkonzeption die öffentliche Aufarbeitung des keltischen Erbes. Dazu gehörte eine umfangreiche Sonderausstellung, die das keltische Leben mitsamt seiner Götterwelt, seiner Ahnenverehrung und seiner heiligen Plätze präsentiert. Die Schau im Konstanzer Archäologischen Landesmuseum inszeniert all diese Aspekte in einer multimedial unterstützten Erlebniswelt, die die Besucher:innen atmosphärisch umfängt und das Thema über Installationen transportiert. Dabei wird auch die mythische Seite des keltischen Kults spürbar und durch Lichtszenen sowie Medienstationen immersiv präsent.

The State of Baden-Württemberg's »Celt Concept« aims to encourage the public to reflect on the region's Celtic heritage. Among the measures was an extensive special exhibition presenting Celtic life, including its pantheon, worship of ancestors and sacred places. The show at the State Archaeological Museum in Constance stages all these aspects in a multimedia encounter that envelops visitors in its atmosphere and conveys the topic by means of installations. In doing so, it succeeds in making the mythical side of Celtic worship palpable as well, using light scenographies and media stations to create an immersive experience.

JURY STATEMENT

Dieses Hörspiel erlaubt nicht nur der jüngeren Generation einen ganz besonderen Zugang zu einem weitgehend unbekannten historischen Thema. Das Format macht Geschichte lebendig, ist barrierefrei und obendrein ressourcenschonend.

Because it provides a very special kind of access to a largely unknown historical reality, the audio drama's appeal is by no means limited to the younger generation. The format brings history to life and is both barrier-free and resource-friendly.

AUFTRAGGEBER/CLIENT
Haus der Geschichte
Baden-Württemberg
Stuttgart

DESIGN
Klangerfinder GmbH & Co KG
Stuttgart

Es ist eine Situation mit historisch brisanten Dimensionen: Nach 1945 arbeiten bei der Stuttgarter Kriminalpolizei 15 ehemalige Opfer des NS-Regimes und 14 ihrer damaligen Verfolger als Kollegen Tür an Tür. Dies belegen Akten, Bewerbungsschreiben und Passbilder. Um das eher spröde Archivmaterial in ein zeitgemäßes Format zu bringen, ersannen die Ausstellungsmacher:innen ein interaktives »True-Crime«-Hörspiel. Die eigens entwickelte Audio-App stellt erzählerisch sowie in Dialogen die Perspektiven der Verfolger und Verfolgten gegenüber, ein dramaturgisch eingesetztes und emotionales Sounddesign hält die Spannungsbögen hoch.

It's a situation with highly charged historical dimensions: after 1945, 15 former victims of the Nazi regime and 14 of their persecutors are now colleagues, working next door to one another as police officers in Stuttgart's Criminal Investigation Department. This is documented by files, application letters and passport photos. In order to convert these rather dry archive materials into a contemporary format, the exhibition makers came up with the idea of a »true crime« audio drama. In narration and dialogue, the specially developed audio app juxtaposes the perspectives of the persecutors with those of the persecuted, underlaid with a dramatically deployed and emotional sound design that keeps up the suspense.

SILVER

ANTHROPOZÄN
AUSSTELLUNGSKONZEPTION
EXHIBITION CONCEPT

> **JURY STATEMENT**
>
> Das Rückgrat der Ausstellung bildet ein Baugerüst, das in bewusstem Kontrast zum historischen Interieur des Museums steht. Diese Einfachheit wird dem großen Thema ebenso gerecht wie der Verzicht auf Verbundmaterialien oder die didaktische Aufbereitung, die auch positive Perspektiven aufzeigt.
>
> The exhibition design centres on a backbone of scaffolding – in deliberate contrast to the historic interior of the museum. This simplicity is just as compatible with the issue at stake as the absence of composite materials and the didactic approach, which includes positive perspectives as well.

AUFTRAGGEBER/CLIENT
Staatliches Museum
für Naturkunde
Stuttgart

DESIGN
Raumhochn
Stuttgart

Die Hybris des Menschen kennt kaum Grenzen, inzwischen formt er die Erde immer weiter um, leider nicht zum Besseren. Der Klimawandel ist nur ein Beispiel für menschengemachte Katastrophen mit weitreichenden Folgen für nachfolgende Generationen, für die Tier- und Pflanzenwelt. 2021 widmete sich die Große Landesausstellung Baden-Württemberg im Naturkundemuseum Stuttgart dem Anthropozän, also dem vom Menschen bestimmten geologischen Zeitalter. Die Schau zeigt die vielen Wirkebenen samt Ursachen, lädt zum Nach- und Umdenken ein und will dennoch positive Impulse für Veränderung geben.

Das Konzept ist – der Logik des Themas entsprechend – zirkulär gedacht und setzt auf die Nachnutzung der Elemente, auf ökologische Materialien und formale Reduktion.

Human hubris seems to know no bounds and is increasingly transforming the Earth – unfortunately, not for the better. Climate change is just one example of the manmade catastrophes that will have far-reaching consequences not just for future generations but for the planet's flora and fauna as well. In 2021, the State of Baden-Württemberg's special exhibition at Stuttgart Museum of Natural History was dedicated to the Anthropocene, i.e. to the geological epoch defined by human impact. The show presents the many levels on which these effects are taking place together with their causes and invites visitors to reflect on and reconsider the way they think, while nevertheless aiming to provide positive impetus for change. In keeping with the logic of the issue it explores, the concept takes a circular approach and is based on the reuse of the exhibition elements, ecological materials and pared-down forms.

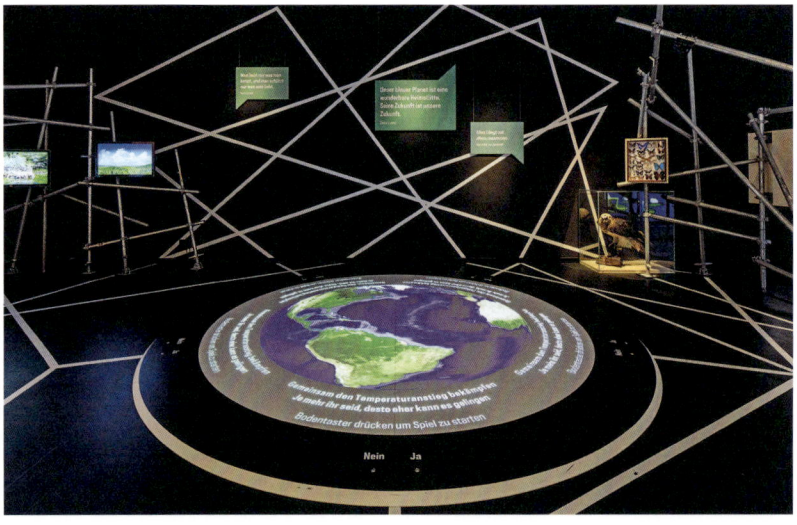

| SPECIAL MENTION | CLIC UP | DISPLAYSYSTEM / DISPLAY SYSTEM |

JURY STATEMENT

Hier wird ein Ausstellungssystem, das sich dank seiner Flexibilität und Einfachheit über Jahrzehnte bewährt hat, mit einem Spannrahmen sinnvoll ergänzt und damit noch intensiver nutzbar. Alte Bestände können so erweitert und noch vielseitiger genutzt werden.

Thanks to its flexibility and simplicity, the display system has stood the test of time and been on the market for decades. Now, the addition of a stretch frame provides a meaningful add-on that permits even greater flexibility. As a result, existing systems can be upgraded to enable even more versatility.

HERSTELLER/MANUFACTURER
Burkhardt Leitner
Modular Spaces GmbH
Stuttgart

DESIGN
Inhouse
Akin Nalca

VERTRIEB/DISTRIBUTOR
Burkhardt Leitner
Modular Spaces GmbH
Stuttgart

Erstmals wurde das Displaysystem im Jahr 1998 vorgestellt; seitdem ist es international im Einsatz. Das System besteht aus filigranen Rundrohren und Knotenwürfeln, die über integrierte Magnete sicher verbunden werden. Nun erhält der Klassiker ein funktionales Upgrade, das ein wichtiges Feature heutiger Präsentationen integriert: die großflächige Bespannung mit textilen Flächen. Das speziell entwickelte Zusatzprofil lässt sich dank des identischen Knotenwürfels exakt in das bewährte Raster einbinden – damit sind auch vorhandene Systeme erweiterbar. Ein selbstschließendes Scharnier ermöglicht den werkzeuglosen Aufbau des Rahmens, die bedruckten Textilelemente werden ebenfalls werkzeuglos in den Rahmen eingehängt und dabei gespannt. Eine Hinterleuchtung ist möglich – und auch eine gleichzeitige Nutzung der Vorder-und Rückseite des Displays.

First launched back in 1998, the display system has been in use across the world ever since. The system consists of slender tubes and connecting cubes that are held securely in place by integrated magnets. The classic has now been given a functional upgrade that integrates an important feature of contemporary presentations: large-format textile coverings. Thanks to its identical connecting cube, the specially developed profile is a perfect fit with the tried-and-tested grid and can therefore be incorporated into existing systems as well. A self-locking hinge means the frame can be set up without tools. The printed textile elements are inserted into the frame and stretched across it – again, no tools required. Backlighting is also possible – and both sides of the display can be used simultaneously.

META

URBANHARBOR

QUARTIERSTRANSFORMATION
URBAN TRANSFORMATION

URBAN HARBOR

QUARTIE
TRANSFO

PUBLIC DESIGN, URBAN DESIGN
PUBLIC DESIGN, URBAN DESIGN

152
153

META
PUBLIC
DESIGN

RMATION

URBANHARBOR

QUARTIERSTRANSFORMATION
URBAN TRANSFORMATION

JURY STATEMENT

Ein visionäres Beispiel für ein neues Verständnis von urbanem Arbeiten. Das ganzheitlich gedachte Konzept ist langfristig angelegt und integriert aktuellste technische Optionen für eine klimaneutrale und nachhaltige Immobiliennutzung, die zudem in direktem Bezug zur Stadtgesellschaft steht.

A visionary example of a new interpretation of work in urban environments. The holistic concept is designed for the long term and integrates state-of-the-art technical options for the climate-neutral and sustainable use of real estate that also has direct links with the local community.

AUFTRAGGEBER/CLIENT
maxmaier urbandevelopment
Ludwigsburg

KONZEPTION/CONCEPT
SFP Architekten GmbH
Stuttgart

Industrielle Standorte stehen hierzulande vor großen Veränderungen – neue Technologien und die Verlagerung der Produktion erfordern andere Gebäudekonzepte. Meist bedeutet dies Abriss sowie Neubau. Der Ludwigsburger »Urbanharbor« hingegen basiert auf der Ertüchtigung ehemaliger Produktionshallen zu flexibel nutzbaren Gebäuden, die Raum für agiles Arbeiten und kreative Workflows bei innovationsfördernder Campus-Atmosphäre ermöglichen.

Alte Hallen, deren einstiger Zweck noch spürbar ist, wurden sukzessive nach dem Prinzip Haus-im-Haus mit neuen, modularen Arbeitsstrukturen aufgeladen und fit für die Zukunft gemacht. Die jüngste Transformation, eine »Hybrid Loop« genannte, 10.000 Quadratmeter große Halle, sticht hierbei mit einem integrativen Energiesystem hervor, das dank Photovoltaik, Wärmepumpe, Vernetzung und Anbindung an lokale Energienetze unterm Strich einen klimapositiven Betrieb ermöglicht.

Industrial sites in this part of the world are facing major changes – new technologies and the relocation of production call for different building concepts. In most cases, that means demolition and new construction. The Urbanharbor in Ludwigsburg, on the other hand, is based on reconditioning former production facilities and turning them into versatile buildings that provide space for agile working and creative workflows in an innovation-friendly campus atmosphere.

Taking the building-within-a-building principle as a starting point, old industrial halls, the former purpose of which is still palpable, were successively equipped with new, modular working structures and made fit for the future. The latest transformation, a 10,000-square-metre hall called the Hybrid Loop, stands out for its integrated energy system which, thanks to photovoltaics, a heat pump, interconnectedness and linkage with the local grid, permits climate-positive operation.

MAX MAIER — INHABER UND FOUNDER, MAXMAIER URBANDEVELOPMENT

»Unser Ziel ist es, Natur, Mensch, Räume und Technik zu einem urbanen Wirkungsgefüge zu vereinen, aus dem das Neue wachsen kann.«

»Our goal is to bring nature, people, spaces and technology together to create an urban ecosystem that the new can grow out of.«

MAX MAIER — INHABER UND FOUNDER, MAXMAIER URBANDEVELOPMENT

→ **Wo liegen die Wurzeln des heutigen Urbanharbor-Areals?**
Die ersten Industrie- und Gewerbebetriebe siedelten sich Ende des 19. Jahrhunderts in der Ludwigsburger Weststadt an, in unmittelbarer Nähe des neu eröffneten Bahnhofs. Viele der Unternehmen, die zunächst als kleine Handwerksbetriebe anfingen, wuchsen im Verlauf des 20. Jahrhunderts zu bedeutenden Großbetrieben mit entsprechenden Produktionshallen.

Eine dieser Produktionsstätten war das Areal der Firma Eisfink. Das Traditionsunternehmen war 1972 von Asperg in die Weststadt gezogen, befand sich aber zehn Jahre später in wirtschaftlichen Schwierigkeiten. Ich stieg dort ein, führte das Unternehmen aus der Krise und fokussierte mich im weiteren Verlauf auf die Entwicklung der Produktionsflächen von Eisfink durch Transformation des Areals von der rein industriellen und gewerblichen Nutzung hin zu Dienstleistungsbetrieben wie Architekturbüros, Marketingagenturen und großflächigem Einzelhandel für die Nahversorgung.

Wie sieht Ihr Konzept des Urbanharbors aus?
Ich habe meinen Fokus bei der Arealentwicklung stets darauf gelegt, Leben und Arbeiten zu vereinen. Der Gedanke war, eine Stadt in der Stadt zu bauen. Dafür braucht es lebendige Gemeinschaftsflächen – deshalb haben wir zu Beginn der Nullerjahre Industriehallen zu Gastronomie- und Eventlocations revitalisiert, wie beispielsweis das Werkcafé und das Alte Werkcafé.

Den nächsten Meilenstein in Richtung Arbeiten und Leben der Zukunft haben wir im November 2016 mit der Eröffnung der Rieber Flagshipkitchen Speisewerk gesetzt. Wir haben die ehemalige Hüller-Hille-Halle zum Dreh- und Angelpunkt des Urbanharbors transformiert – heute bieten wir hier CO_2-neutrale Arbeitsplätze für junge, kreative Start-ups, etablierte Unternehmen sowie Gastronomie und Handel.

Wie kam es zur Idee dieser Transformation eines industriellen Standorts?
Die Vision hinter Maxmaier Urbandevelopment ist, den Bezug zur industriellen Historie des Areals zu erhalten. Die Architektur des Ortes bewahrt stets die Energie der Vergangenheit, das macht die alten Industriebrachen zu Räumen des Lebens. Leben ist vielschichtig, deshalb verfolge ich mit Urbanharbor die Intention, das Areal zu einer eigenen Stadt, mit einem eigenen Ökosystem von Unternehmen, Gastronomie, Einzelhandel und Unterhaltung aufzubauen. Unser Ziel ist es, Natur, Mensch, Räume und Technik zu einem urbanen Wirkungsgefüge zu vereinen, aus dem das Neue wachsen kann.

Eine Transformation braucht auch passende Planer – wie haben Sie das Büro SFP gefunden?
Bereits 2016/2017 sind wir über unseren Mieter Grow, eine Start-up-Tochter von Bosch, mit SFP Architekten zusammengekommen. Für uns war diese Vernetzung eine Bereicherung im Hinblick auf neue Denkweisen und Konzepte. Es folgte dann die partnerschaftliche Projektzusammenarbeit für das CO_2-neutrale Objekt Hybrid Loop, welches seit Mitte 2021 fertiggestellt ist.

Wie wichtig ist die kommunale Unterstützung des Prozesses?
Wir haben einen guten Kontakt mit der Stadt und den zuständigen Verantwortlichen. Im Rahmen des Machbaren wird stets ein Konsens gefunden, denn unser Stadtquartier ist für die Stadt Ludwigsburg wirtschaftlich, sozial und ökologisch gesehen ein Mehrwert.

Was würden Sie Unternehmen an die Hand geben, die einen ähnlichen Weg andenken?
Zuerst muss man sich perspektivisch fragen: Was ist der Kontext und was kann daraus in der Nutzung und Funktion für Menschen entstehen? Denn erst, wenn für die Menschen Funktion und Nutzung unter ökologischen Aspekten reflektiert und bewertet sind, kann mit der entsprechenden Transformation begonnen werden. Das bedeutet, dass Architektur und Design stets der Funktion und Nutzung folgen müssen.

Sie haben einen illustren Nutzermix – wie wichtig ist die Durchmischung?
Existenziell, denn die Stadt der Zukunft muss neu gedacht werden, von der Bevölkerungsstruktur, der Geschichte, von den Gebäuden sowie von der Wirtschaftsstruktur her. Jede Stadt hat ihre einzigartige DNA, je vielfältiger der Kreis der Nutzenden und die Bespielung ist, umso höher die Diversität.

Die Stadtwerke spielen eine wichtige Rolle im Energiekonzept. Welche genau?
Die Stadtwerke Ludwigsburg sind für uns bezüglich Energiegewinnung, -speicherung und -verteilung unverzichtbar. Zudem gehen sie mit uns gemeinsam in eine nachhaltige Zukunft mit erneuerbaren Energien sowie digitalen Plattformlösungen, die bei uns auf dem Areal pilotiert werden.

Wie wird sich das Areal weiter entwickeln?
Das Zusammenleben und Arbeiten in Städten wird sich grundlegend verändern, es wird verschmelzen. Es ist der Mensch, der bei dem Wandel der Arbeitswelt für uns im Zentrum steht. Seine Bedürfnisse und Wünsche sollten bei der Gestaltung seines Arbeitsumfelds berücksichtigt werden.

Die aktuell wichtigste Entwicklung für unser Areal hinsichtlich der Energieversorgung ist die CO_2-Neutralität, die wir uns bis 2030 für das komplette Areal zum Ziel gesetzt haben.

Urbandevelopment initiiert architektonische Raum-, Immobilien- und Stadtentwicklung für das Ludwigsburger Areal Urbanharbor und die dort tätigen Menschen. Das alte Industriegebiet mit rund 200.000 Quadratmetern vereint architektonische, ökonomische, ökologische und soziale Werte zu einer vernetzten und klimaneutralen Stadt der Zukunft.

https://urbanharbor.com

MAX MAIER

OWNER AND FOUNDER, MAXMAIER URBANDEVELOPMENT

→ Can you tell us about the roots of what is now the Urbanharbor complex?

The first industrial and commercial enterprises set up business in the Weststadt area of Ludwigsburg in the late 19th century, in the direct vicinity of the newly opened station. In the course of the 20th century, a lot of the companies that had started out as small workshops grew into major firms with correspondingly large production halls.

One of those production facilities was a complex belonging to a company called Eisfink. The old-established company had moved from Asperg to the Weststadt district in 1972, but 10 years later it found itself in financial difficulties. I came on board, led the company out of crisis and then focused on developing Eisfink's production facilities by transforming the complex, moving away from purely industrial and commercial use and towards service providers like architectural firms, marketing agencies and large-scale retail for the local population.

Can you outline your concept for the Urbanharbor?

When I was developing the site, I always focused on combining life and work. The idea was to build a city within the city. That calls for lively communal areas – which is why, in the early 2000s, we started revitalising industrial facilities by turning them into gastronomic and event locations, like the Werkcafé and the Alte Werkcafé.

The next milestone on the road to future-fit working and living came in 2016 when we opened the Speisewerk, which is the Rieber flagship kitchen. We transformed the former Hüller Hille hall into the focal point of the Urbanharbor – today we offer carbon-neutral workspaces for young, creative startups, established firms, gastronomy and retail.

How did you hit on the idea of transforming an industrial site?

The vision behind Max Maier Urbandevelopment is to preserve the link with the site's industrial history. The architecture of a place always conserves the energy of the past, and that makes old industrial wastelands spaces of life. Life is multilayered – that's why, with Urbanharbor, my intention is to turn the complex into a city in its own right, with its own ecosystem of businesses, gastronomy, retail and entertainment. Our goal is to bring nature, people, spaces and technology together to create an urban ecosystem that the new can grow out of.

A transformation calls for the right planners – how did you find SFP, the architectural firm that you collaborated with?

We first got together with SFP back in 2016/2017 via our tenant Grow, a startup subsidiary of Bosch. For us, this kind of networking was definitely a bonus in terms of new ways of thinking and concepts. Then we partnered with them on the carbon-neutral Hybrid Loop project, which was completed in mid-2021.

How important is the local community's support for the process?

We have a good relationship with the city and the relevant authorities. Within the bounds of possibility, we always find a consensus because our development is a value-add for the city of Ludwigsburg – not just in an economic sense, but socially and ecologically as well.

What advice would you give companies who are considering a similar path?

First you have to think about it in perspective: what's the context and what can its use and function do for people? Because you can't get the transformation underway until you've considered and evaluated its function and use for people from an ecological standpoint. That means the architecture and design always have to be derived from the function and use.

You've got an illustrious mix of users – how important is the right mixture?

It's existential, because the city of the future has to be rethought with regard to its population structure, history, buildings and economic structure. Every city has its unique DNA, and the more varied the mix of users and the »occupancy« is, the greater the diversity.

The municipal utilities play an important role in the energy concept. What exactly does that role consist of?

Ludwigsburg's municipal utilities are indispensable for us when it comes to energy production, storage and distribution. What's more, they're joining us on the path to a sustainable future with renewable energies and digital platform solutions that are being piloted on our premises.

How will the complex develop from this point on?

The way we live and work in cities will change fundamentally: the two spheres will merge. For us, the transformation of the world of work centres on people. Their wants and needs should be taken into account when designing their work environments.

In terms of energy supply, the most important development for our complex right now is carbon neutrality: we've set ourselves the goal of making the entire complex carbon-neutral by 2030.

Urbandevelopment initiates architectural space, real estate and urban development for Ludwigsburg's Urbanharbor hub and its people. On an area of approx. 200,000 square metres, the old industrial park blends architectural, economic, ecological and social values into a connected and climate-neutral city of the future.

https://urbanharbor.com

1 Vorher – Nachher
Before – After

2 Vorher – Nachher
Before – After

1 → SEITE/PAGE
162–167

2 → SEITE/PAGE
168, 170

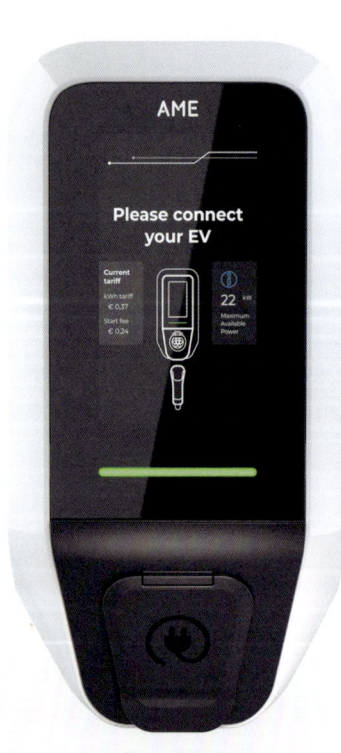

3 → SEITE/PAGE
169, 171

MOBILITY

GOLD:
1 **CL3810**
Recaro Aircraft Seating GmbH & Co. KG
Schwäbisch Hall

SILVER:
2 **BRUNO**
Kuckoo Camper GmbH & Co.KG
Löchgau

SPECIAL MENTION:
3 **CHARLOT**
AME B.V.
Eindhoven
Niederlande/The Netherlands

Die Mobilität ist ein Schlüsselthema der Moderne – und eines, das vor großen Transformationen steht. Zugleich diversifizieren sich die Mobilitätsangebote immer weiter, einschließlich spezifischer Detaillösungen für Services, Wartung oder Individualisierung. Mobility bleibt spannend.

Mobility is a key theme of the modern age – and one that is facing major transformations. At the same time, mobility offerings are becoming increasingly diversified, including specific detail solutions for services, maintenance or customisation. Mobility remains a fascinating field.

12

GOLD CL3810 LANGSTRECKEN-FLUGZEUGSITZ
LONG-HAUL AIRCRAFT SEAT

MOBILITY
MOBILITY

162
163

FOCUS GOLD

CL3810 LANGSTRECKEN-FLUGZEUGSITZ / LONG-HAUL AIRCRAFT SEAT

JURY STATEMENT

Hoher Komfort bei schlanker und gewichtsoptimierter Konstruktion – das muss kein Widerspruch sein. Der Langstreckensitz bietet zahlreiche durchdachte Details, etwa die Magazintasche, die schlanken Armlehnen und die Nackenstütze. Die Qualitätsanmutung stimmt ebenso wie die sehr aufgeräumt gestaltete Rückseite, die große Displays aufnehmen kann.

There's no reason why a high level of comfort and a lean, weight-optimised design need to be mutually exclusive. The long-haul seat features numerous well thought-out details, including the literature pocket, the slender armrests and the neck support. The impression of quality is just as compelling as the uncluttered design of the back, which is able to accommodate large screens as a result.

HERSTELLER/MANUFACTURER
Recaro Aircraft Seating GmbH & Co. KG
Schwäbisch Hall

DESIGN
Inhouse

VERTRIEB/DISTRIBUTOR
Recaro Aircraft Seating GmbH & Co. KG
Schwäbisch Hall

Das Streben nach weniger Gewicht begleitet die Luftfahrt seit Anbeginn. Recaro arbeitet daher an der Verschlankung und Optimierung der Passagiersitze, ohne den Komfort zu reduzieren. Der neue Sitz wiegt – je nach Ausstattungsdetails – zwischen 12,6 und 13,3 Kilogramm, er ist damit 10 bis 15 Prozent leichter als frühere Sitze. Ein Jet mit 280 Plätzen wird so um rund 420 Kilogramm leichter – ist er voll ausgelastet unterwegs, so reduziert sich die CO_2-Emission pro Platz und Jahr um 108 Kilogramm.

Das Komfortthema wird ebenfalls relevanter, die geopolitische Lage erfordert Umwege und damit vermehrt extreme Langstreckenflüge. Der Sitz bietet daher mehr Beinfreiheit und eine ergonomischere Ruheposition: Wird die Rückenlehne geneigt, hebt sich die Vorderseite der CFK-Sitzschale. Eine mehrfach verstellbare Kopfstütze mit optionalem Nackenkissen sowie große, seitliche Flügel im Kopfbereich sind ebenfalls neu.

Less weight is a goal the aviation industry has aspired to since its very beginnings. Recaro's objective is therefore to slenderise and optimise passenger seats without sacrificing comfort. Depending on the features selected, the new seat weighs between 12.6 and 13.3 kilograms and is therefore 10 to 15% lighter than earlier models. That means a weight saving of approx. 420 kilograms for a 280-seat jet, reducing CO_2 emissions by 108 kilograms per seat and year if it operates at full capacity.

Comfort is also becoming increasingly relevant: the current geopolitical situation is making detours necessary, resulting in more and more extremely long flights. The seat therefore provides more legroom and a more ergonomic resting position: if the backrest is declined, the front of the CFRP seat pan tilts up. The new features also include a multi-adjustable headrest with optional neck support, as well as large wings on the sides of the head section.

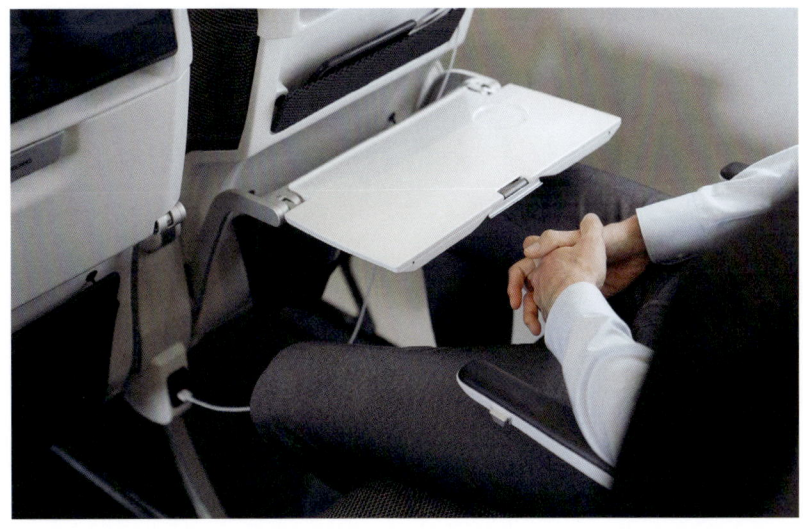

STEFAN BERROTH — PRODUKTMANAGER, RECARO AIRCRAFT SEATING GMBH & CO. KG

»Vom ersten Entwurf bis zum Bau der Produktionswerkzeuge arbeiten Industriedesign, Engineering und Ergonomie in interdisziplinären Teams eng zusammen.«

»From the first sketch all the way to tooling, we bring industrial design, engineering and ergonomics together to collaborate closely in interdisciplinary teams.«

STEFAN BERROTH
PRODUCT MANAGER, RECARO AIRCRAFT SEATING GMBH & CO. KG

→ **Warum brauchen wir für Langstreckenflüge besondere Sitzkonzepte?**

Bei einer Flugdauer von bis zu 16 Stunden, oft auch bei Nachtflügen, spielt das Ausruhen und Schlafen im Sitz eine wichtige Rolle. Das unterstützen beispielsweise verstellbare Kopfstützen mit Seiten- und Nackenstütze, Sitzkissen sowie Sitzstrukturen mit optimaler ergonomischer Unterstützung für aufrechtes Sitzen und für besten Komfort bei einer um 30 Grad geneigten Rückenlehne. Außerdem erwartet der Passagier auf längeren Flügen Unterhaltung: Daher haben wir modernste Bildschirmtechnik sowie Lade- und Haltestationen für die eigenen Smart Devices in den Sitz integriert.

Wo wird denn der Komfort des CL3810 spürbar?

Größere Passagiere nehmen zuerst den um 25 Millimeter erweiterten Knieraum wahr. Verglichen mit seinem Vorgängermodell reduziert der CL3810 auch den Druck auf die Oberschenkel in entspannter, gestreckter Position dank der neuen Sitzschale mit flexibler Vorderkante und Mehrzonenschaum. Auch die bereits erwähnte Nackenstütze wurde verbessert. Für den visuellen Komfort seien die Minimierung der Kunststoffbereiche und die Vergrößerung der Textilbereiche auf der Rückseite und an den Seiten des Sitzes genannt. Und schließlich hat das Design der Bedieneinheiten für die Bordunterhaltung (IFE) erheblichen Einfluss auf das Platzangebot für die Passagiere.

Wie entwickelt man ein ergonomisches Produkt für einen internationalen Markt?

Vor Beginn der Entwicklung definieren wir viele Sitzparameter und -abmessungen, die während des Entwicklungsprozesses kontinuierlich validiert und evaluiert werden. Daneben setzen wir auf das RAMSIS-Modell, auf virtuelle Nachbildungen verschiedener Passagiertypologien und natürlich auf Tests mit individuellen Testpersonen. Wir versuchen immer, unsere Sitze für eine internationale Probandengruppe anzupassen – von einer Frau mit 1,55 Metern bis zu einem Mann mit 1,92 Metern Größe. Dabei berücksichtigen wir auch die unterschiedlichen Körpermaße etwa von asiatischen und europäischen Probanden.

Wie eng arbeiten Technik und Design bei der Entwicklung zusammen?

Vom ersten Entwurf bis zum Bau der Produktionswerkzeuge arbeiten Industriedesign, Engineering und Ergonomie in interdisziplinären Teams eng zusammen. Wenn wir Konzepte entwickeln und Produkte entwerfen, dann greifen wir auch auf Informationen von Kunden, unseres Verkaufs und Produktmanagements, auf Engineering, Fertigung, Modellbau sowie Materialspezialisten zurück. Gewichte und Kosten spielen eine zentrale Rolle, genauso die räumlichen Platzverhältnisse der Passagierkabinen. Die ermittelt beispielsweise unser Digital Mock-up-Team und stellt sie uns zur Verfügung.

> Recaro Aircraft Seating ist ein weltweiter Anbieter von Premium-Flugzeugsitzen. Der Marktführer für Economy Class Sitze steht für Produktinnovationen und preisgekrönten Kundenservice gemäß dem Motto »driving comfort in the sky«. Neben dem Ausbau seiner globalen Präsenz setzt das Familienunternehmen Recaro auf Zuverlässigkeit, Effizienz, Nachhaltigkeit und den Zuwachs an Mitarbeiter:innen.
>
> www.recaro-as.com

→ **Why do we need special seating concepts for long-haul flights?**

When you're on a flight that lasts up to 16 hours, and often on night flights too, being able to rest and sleep in your seat plays an important role. So it can really help if it's equipped with things like adjustable headrests with side and neck supports, or seat cushions and seat structures that provide optimal ergonomic support when sitting upright and maximum comfort when the backrest is reclined at a 30-degree angle. On top of that, passengers on longer flights expect entertainment too, which is why we've integrated state-of-the-art screen technology into the seat, as well as chargers and holders for the occupant's own smart devices.

How does the extra comfort the CL3810 provides make itself felt?

The first thing taller passengers notice is the additional 25 millimetres of kneeroom. And thanks to the new seat pan with a flexible front edge and multizone foam, the CL3810 also reduces the pressure on the thighs when they're in a relaxed, extended position as compared to the predecessor model. The neck support I mentioned just now has also been improved. As for visual comfort, we've minimised the plastic surfaces and enlarged the textile sections on the back and sides of the seat. And last but not least, the design of the control units for the in-flight entertainment has a considerable influence on the space available for the passengers.

How do you go about developing an ergonomic product for an international market?

Before development starts we define a lot of seat parameters and dimensions that are then continuously validated and evaluated during the development process. In addition, we rely on the RAMSIS manikin, virtual simulations of various passenger typologies and of course tests with real-life people. We always try to adapt our seats for an international group of test subjects – from a 1.55-metre-tall woman all the way to a 1.92-metre tall man. We also take the different body dimensions of e.g. Asian and European test subjects into account.

How closely do engineering and design collaborate when it comes to development?

From the first sketch all the way to tooling, we bring industrial design, engineering and ergonomics together to collaborate closely in interdisciplinary teams. When we're developing concepts and designing products, we also fall back on information from clients and our own sales and product management teams, as well as input from engineering, production, model-making and materials specialists. Weight and cost play a key role, as does the space in the passenger cabins – not just how much of it is available, but how it's allocated. That's something our digital mockup team works out and makes available to us.

> Recaro Aircraft Seating is a global supplier of premium aircraft seats. The market leader in economy class seating stands for product innovations and award-winning customer service that lives up to its motto: »driving comfort in the sky«. In addition to expanding its global presence, the family-owned company is committed to reliability, efficiency, sustainability and employee growth.
>
> www.recaro-as.com

SILVER	BRUNO	MINI-CAMPER
	→ SEITE/PAGE 170	MINI CAMPING TRAILER

SPECIAL MENTION | CHARLOT → SEITE/PAGE 171 | LADESTATION EV CHARGER

SILVER

BRUNO

MINI-CAMPER
MINI CAMPING TRAILER

> **JURY STATEMENT**
>
> Der kleine Wohnanhänger bringt neue Ideen und ein neues Werkstoff-Verständnis ins Spiel, weil für seinen Bau statt Kompositplatten stabiles Schichtholz verwendet wird. Das Konzept ist funktional absolut durchdacht und wartet mit liebevoll gestalteten Details auf. Hier spürt man, dass die Macher auch selbst Nutzer sind.
>
> Because it's made of sturdy plywood rather than composite panels, the little camping trailer brings new ideas and a new approach to materials into play. Besides being totally compelling in terms of functionality, the concept also features some lovingly designed details. You can tell that its makers are users as well.

HERSTELLER/MANUFACTURER
Kuckoo Camper GmbH & Co.KG
Löchgau

DESIGN
Inhouse

VERTRIEB/DISTRIBUTOR
Kuckoo Camper GmbH & Co.KG
Löchgau

Seit der Pandemie kennt das Wachstum in der Camper- und Wohnmobilbranche nur eine Richtung: nach oben. In Zeiten eingeschränkter Reise- und Übernachtungsmöglichkeiten scheint das eigene Heim auf Rädern eine vielversprechende Alternative zu sein. Nachhaltig ist das nicht immer – hier versucht der Mini-Camper Maßstäbe zu setzen. Der gebremste Anhänger aus Mehrschichtholz mit PU-Versiegelung ist klein, leicht und kompakt, trotzdem bietet sein durchdachtes Innenleben Platz für eine Schlafstatt, Schränke, Ablagen und eine ausziehbare Küchenschublade. Passendes Zubehör wie etwa Dach- oder Fahrradträger, Solaranlage oder Markise ist optional erhältlich. Nachhaltigkeit ist auch in der Firmenphilosophie verankert: Für jeden verkauften Camper werden fünf neue Bäume gepflanzt.

Since the pandemic, the camper and caravan industry has only been trending in one direction: upwards. In times when travel and accommodation options are limited, a temporary home on wheels seems like a promising alternative. But it's not always a sustainable option – which is why the mini camping trailer is trying to set new standards. Made of PU-sealed plywood, the braked trailer is small, light and compact, yet its cleverly designed interior still provides enough space for a bed, cupboards, shelves and a pullout kitchen drawer. Optional accessories including a roof rack, bike rack, solar panels and an awning are also available. Sustainability is firmly anchored in the company's philosophy as well: five new trees are planted for every trailer sold.

SPECIAL MENTION — CHARLOT — LADESTATION / EV CHARGER

JURY STATEMENT

Ein formal sehr schlüssiges und zeitgemäßes Produkt, dessen großes Display gut in die Front integriert wurde und das schnell erfassbare Interface in den Vordergrund rückt. Die Ladestation kann sich in unterschiedliche Interiors einfügen.

A very consistently designed and contemporary product. The large display is well integrated and focuses attention on the instantly comprehensible interface, while the charger's understated appearance will blend in with different interiors.

HERSTELLER/MANUFACTURER
AME B.V.
Eindhoven
Niederlande/The Netherlands

DESIGN
GBO Innovation Makers
Helmond
Niederlande/The Netherlands

VERTRIEB/DISTRIBUTOR
AME B.V.
Eindhoven
Niederlande/The Netherlands

Soll die Energiewende gelingen, braucht es viele Ladestationen für Elektrofahrzeuge, auch im privaten Umfeld. Dafür ist dieses schlanke Gerät gedacht: Es lässt sich in das Smart Home integrieren, unterstützt bidirektionales Laden und trägt so zur Stabilisierung der Stromnetze bei. Maximal steht eine Wechselstromleistung von 22 Kilowatt bereit, ein Schnellverbindungssystem ermöglicht die Montage an der Wand oder an einem Mast. Die leicht schräg ansetzende Buchse reduziert den Platzbedarf des meist großen Steckers und das große Display in Kombination mit der LED-Statusleiste zeigt auf einen Blick, wie es aktuell um die Ladesituation steht.

If the energy transition is to succeed, a great many EV chargers will be required – in private settings too. That is precisely what this slender device is intended for: it can be integrated with the smart home, is equipped with bidirectional charging technology and thus helps stabilise the grid. The charger supplies maximum AC power of 22 kilowatts and can be attached to a wall or mast thanks to a quick mounting system. The socket is set at a slight angle to reduce the amount of space taken up by the generally large plug, while the combination of a large display and LED progress bar indicates the current charging status at a glance.

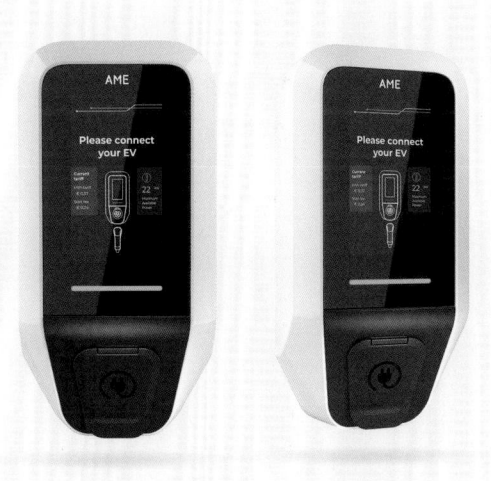

IRMY WILMS-HAVERKAMP **HAVERKAMP INTERIOR DESIGN | KÖHLER WILMS, HERFORD**

»Nachhaltiges Design bedeutet, ein Produkt auch vom Ende her zu denken. Weil es dabei um Wertschöpfungsketten und Kreislaufwirtschaft geht, ist es wichtig, interdisziplinär zu arbeiten.«

»Sustainable design means considering a product from the end as well. Because it has to do with value chains and the circular economy, it's important to work across disciplines.«

Irmy Wilms-Haverkamp ist Mitbegründerin von Köhler Wilms Produktdesign und Partnerin von Haverkamp Interior Design. Schwerpunkte des Herforder Studios sind einerseits strategische Designlösungen für den Konsumgütermarkt, andererseits individuelles Interior Design.
 Die Entwürfe von Köhler Wilms wurden mit zahlreichen Preisen ausgezeichnet und zeichnen sich durch ein umfassendes Verständnis von Zielgruppe und Marke aus.
 Irmy Wilms-Haverkamp studierte Industriedesign an der Universität der Künste in Berlin und lehrte dort nach ihrem Abschluss.

Irmy Wilms-Haverkamp is co-founder of Köhler Wilms Produktdesign and a partner at Haverkamp Interior Design. The Herford-based studio's main areas of focus are strategic design solutions for the consumer goods market on the one hand and individual interior design on the other.
 Köhler Wilms' designs have won numerous awards and stand out for their comprehensive understanding of the target group and brand.
 Irmy Wilms-Haverkamp studied industrial design at Berlin University of the Arts, where she also taught after graduating.

http://www.koehlerwilms.de
http://www.h-id.com

http://www.koehlerwilms.de
http://www.h-id.com

1 → SEITE/PAGE 176–181

2 → SEITE/PAGE 182, 183

SERVICE DESIGN
SERVICE DESIGN

GOLD:
1 **CONTROLLER SERIE 500**
Nabertherm GmbH
Lilienthal

SILVER:
2 **TOURGUIDE 2**
Meder CommTech GmbH
Singen

Die Digitalisierung von Produkten und Dienstleistungen schreitet voran – die nutzerzentrierte Gestaltung von Interfaces und Bedienlogiken sorgt dafür, dass die Services in ihrer Tiefe und Funktionalität zu intuitiven Tools werden, die sowohl im virtuellen wie realen Kontext ihren eigentlichen Wert entfalten können.

The digitalisation of products and services is progressing – the user-centric design of interfaces and operating logics makes the depth and functionality of services accessible by turning them into intuitive tools that can demonstrate their true value in both a real and virtual context.

13

CONTROL SERIE 50 OFENSTEUERUNG

GOLD CONTROLLER SERIE 500 OFENSTEUERUNG KILN CONTROLLER

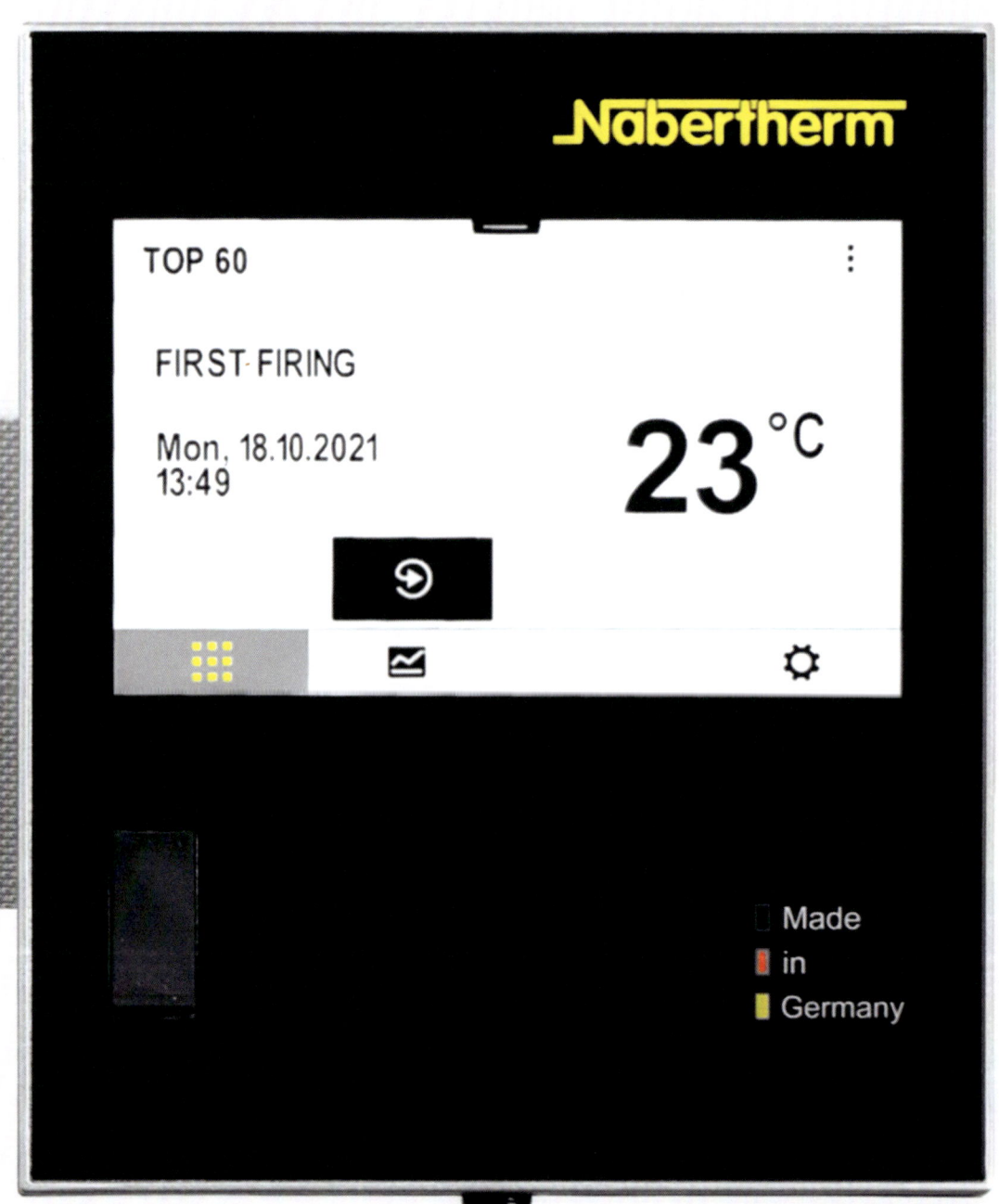

GOLD

CONTROLLER SERIE 500
OFENSTEUERUNG / KILN CONTROLLER

JURY STATEMENT

Ein sehr gutes User Interface, das komplexe Vorgänge schnell erfassbar und transparent darstellt. Parameter lassen sich individuell variieren und festschreiben. Das modulare Screendesign kann auf weitere Produkte übertragen und jeweils erweitert werden. Positiv anzumerken ist auch der Ansatz, alte Screens gegen neue auszutauschen, den Ofen also upgraden zu können.

A very good user interface that depicts complex processes in a readily comprehensible and transparent way. The parameters can be varied and defined to suit individual needs. The modular screen design can be transferred to other products and expanded as required. The approach of replacing old screens with new ones and thus being able to upgrade the kiln is commendable.

HERSTELLER/MANUFACTURER
Nabertherm GmbH
Lilienthal

DESIGN
Zweigrad GmbH & Co. KG
Hamburg

VERTRIEB/DISTRIBUTOR
Nabertherm GmbH
Lilienthal

Brennöfen für keramische Produkte benötigen eine präzise Regelung und fahren je nach Material unterschiedliche, teils komplexe Brennprogramme. Mit dem neuen Interfacesystem lassen sich diese Prozesse exakt aktivieren, individuell anpassen sowie editieren. Auch visualisiert das User Interface mit seinen vielen Ebenen den jeweiligen Brennstatus mit den wichtigsten Parametern. Das Design der Steuerung basiert auf einer engen Kooperation mit den Nutzergruppen, integriert das Markenbild des mittelständischen Herstellers, vereinfacht die professionelle Ofennutzung und verbessert so die Produktqualität. Das System ist modular konzipiert und damit für neue Funktionen, Anforderungen oder Displaytypen offen. Bislang verwendete Segmentanzeigen lassen sich gegen neue Touchscreens tauschen – oder durch externe Displays ergänzen. Für die Fernüberwachung der Prozesse steht eine Smartphone-App bereit.

Kilns for ceramic products require precise control and run different, sometimes complex firing programs depending on the material. With the new interface system, these processes can be activated with precision, adapted to individual requirements and edited. The multilevel user interface also visualises the respective firing status with the most important parameters. The design of the controller is based on close collaboration with users' groups, integrates the manufacturer's branding, simplifies professional use of the kiln and thus improves product quality. The modular concept behind the system means it is open to new functions, requirements or display types. Existing segment displays can be exchanged for new touchscreens – or supplemented with external displays. A smartphone app can be used to monitor the processes remotely.

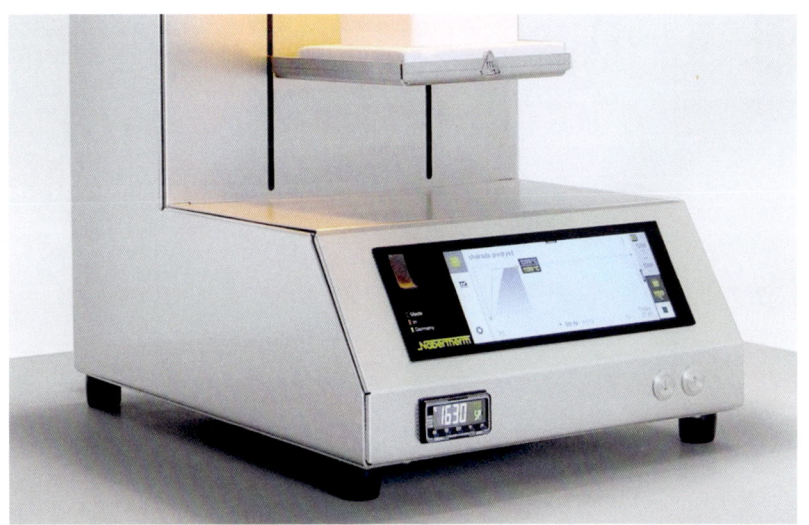

CLAUDIA S. FRIEDRICH   GESCHÄFTSFÜHRERIN INTERFACE DESIGN, ZWEIGRAD DESIGN

»Die Brennprozesse sind recht komplex, also mussten wir zunächst die Abläufe verstehen.«

»The firing processes are quite complex, so to begin with we had to make sure we understood them.«

CLAUDIA S. FRIEDRICH — MANAGING DIRECTOR INTERFACE DESIGN, ZWEIGRAD DESIGN

→ **Die Entwicklung einer Steuerung ist komplex – was war der Anlass für dieses Projekt?**
Es gab mehrere Gründe – erstens ging es darum, den Qualitätsanspruch der Geräte auch auf deren Steuerung zu übertragen. Zweitens standen Digitalisierung und Vernetzung auf der Liste und drittens wollte man dem Wettbewerb einen Schritt voraus sein.

Mit welchen Herausforderungen wurden Sie bei der Interface-Entwicklung konfrontiert?
Die Brennprozesse sind recht komplex, also mussten wir zunächst die Abläufe verstehen und auch, welche Messwerte welche Relevanz haben. Auch die Heterogenität der Nutzergruppen war herausfordernd für uns. Im Keramik-Bereich arbeiten sowohl Profis als auch Hobby-Keramiker:innen mit dem Produkt. Im industriellen Bereich und in Dentallaboren sind primär Profis am Werk. Letztlich sollte es möglich sein, die Brennprogramme von allen Gruppen selbst editieren und anpassen zu lassen. Und es galt, modular zu denken, damit Interface und Steuerung offen für künftige Erweiterungen sind. Klar war immer, dass die Gestaltungsqualität auf dem Niveau von Smartphone-Anwendungen sein soll, allerdings auf Industriedisplays laufen muss.

Messwerte sind spröde – wie macht man daraus visuell ansprechende Darstellungen?
Wir haben zunächst die Interaktion und Informationshierarchie analysiert, dann die Bedienebenen definiert und parallel dazu die visuelle Erscheinung entwickelt. Die war vor allem vom Markenbild des Unternehmens und der Relevanz der Werte bestimmt. Also gaben wir dem Unternehmens-Gelb die oberste Priorität und ordneten Gelb den relevantesten Informationen zu. Als Ergebnis haben wir eine Farbigkeit aus Gelb und Schwarz, also einen sehr hohen Kontrast, damit werden die Angaben auch aus der Entfernung sofort erfassbar.

Sicher haben Sie das intensiv mit den Nutzer:innen getestet.
Natürlich, aber das war nicht ganz einfach, weil das Projekt mitten in der Pandemie begonnen wurde und wir in die Remote-Arbeitsweise wechseln mussten. Wir konnten also keine klassischen Userrunden durchführen, dafür aber ganz eng den Service und den Vertrieb des Unternehmens einbinden, die den direkten und kontinuierlichen Kontakt zu Nutzer:innen haben. Von dort aus ging es dann oft in den konkreten Abgleich mit Kund:innen, deren Feedback agil ins Prototyping einfloss. Die Stakeholder waren also immer wieder involviert.

Interessant ist auch die bereits erwähnte Display-Lösung.
Beim Start des Projektes hatten die Geräte entweder ein integriertes oder ein additives Display mit Segmentanzeige. In beiden Fällen konnten wir die alten Anzeigen gegen neue, farbige Touch-Industriedisplays austauschen, die robust und zudem auch langfristig verfügbar sind. Wir mussten also keine Änderung an der Hardware der Öfen selbst vornehmen.

---

zweigrad Industrial Design mit Sitz in Hamburg ist ein Designstudio mit dem Fokus auf technische Produktsysteme und der Gestaltung von Schnittstellen zwischen Mensch und Maschine. Zum Team gehören Industrie- und Interface Designer:innen, Ingenieur:innen und Expert:innen für strategische Designführung.

www.zweigrad.de

→ **The development of a controller is complex – what was the motivation behind the project?**
There were several reasons – firstly, it was a question of transferring the quality standard of the equipment to its controls. Secondly, digitalisation and connectivity were part of the agenda too, and thirdly the desire to be one step ahead of the competition.

What challenges did the development of the interface confront you with?
The firing processes are quite complex, so to begin with we had to make sure we understood them, as well as understanding the relevance of the individual readings. The fact that the user groups are so diverse was also challenging. In the ceramics sector, both professionals and amateurs work with the product. In the case of industrial firms and dental labs, it's mainly professionals. Ultimately, the goal was to ensure that all user groups would be able to edit and adapt the firing programmes. And we had to think modularly too, so that the interface and controller can be used for future additions to the product portfolio. It was always clear that the design quality of the interface should be on the same level as smartphone apps, even though it has to run on industrial displays.

Readings aren't exactly exciting – how do you display them in a visually appealing way?
To begin with we analysed the interaction and the information hierarchy, then we defined the control levels and developed the visual implementation parallel to that. First and foremost, the look was determined by the company's brand identity and the relevance of the settings. That's why we gave the corporate yellow top priority and assigned yellow to the most relevant information. The result is a combination of yellow and black, i.e. a very strong contrast – and that means the data is immediately legible, even from a distance.

I imagine you tested that very thoroughly with users?
Yes, of course, but it wasn't easy because we started the project right in the middle of the pandemic and had to switch to remote working. That meant we couldn't conduct classic usability workshops, so we worked very closely with the client's service and sales divisions, who have direct and continuous contact with the users. From there, the next step was often a concrete test with customers, whose feedback was incorporated into the agile prototyping process. So the stakeholders were involved at multiple stages.

The display solution you mentioned earlier on is particularly interesting.
When we started on the project the devices still had a segment display, sometimes integrated, sometimes additive. In both cases we were able to replace the old displays with new, industrial-grade colour touchscreens, which are robust and will be available in the long term. That meant we didn't have to make any changes to the hardware or the furnaces and kilns themselves.

---

zweigrad Industrial Design is a Hamburg-based design studio that focuses on technical product systems and the design of human-machine interfaces. The team includes industrial and interface designers, engineers and strategic design management experts.

www.zweigrad.de

SILVER **TOURGUIDE 2** **AUDIOGUIDE**
AUDIO GUIDE
→ SEITE/PAGE
183

JURY STATEMENT

Eine sehr schöne Idee, die neuartige Rückfragefunktion durch das Drehen des Geräts zu aktivieren. Auch die große Kanalanzeige zeigt, dass hier nutzerorientiert gedacht wurde. Zugleich ist das smart anmutende Gerät auf schnelle Reparierbarkeit hin optimiert.

Activating the innovative talkback function by simply turning the device upside down is a very appealing idea. The large channel display likewise underscores the user-oriented approach. At the same time, the smart-looking device has been optimised for ease of repair.

HERSTELLER/MANUFACTURER
Meder CommTech GmbH
Singen

DESIGN
Solidfluid Produktdesign
Konstanz/Constance

VERTRIEB/DISTRIBUTOR
Meder CommTech GmbH
Singen

Bei Stadtrundgängen, Werks- oder Museumsführungen gehören Audio-Übertragungssysteme fast schon zum Standard. Die Kommunikation läuft aber nur in eine Richtung, also vom Guide zu den Teilnehmenden, Rückfragen sind nicht vorgesehen. Das ändert sich mit dem neu entwickelten Audioguide, dessen sogenannte Talkback-Funktion sogar Fragen an die gesamte Gruppe erlaubt. Aktiviert wird diese Funktion durch das Umdrehen des schlanken Gerätes, das dann den Charakter eines Stabmikrofons bekommt.

Das Gerätevolumen sowie die weichen Konturen orientieren sich an der Handhabbarkeit für alle Altersgruppen. Das Display mit der Kanalanzeige befindet sich hinter einer dunklen Fläche, die in die Oberseite mit der Kopfhörerbuchse übergeht. Das Gehäuse ist so konzipiert, dass es bei Beschädigungen schnell ersetzbar ist und eine gute Zugänglichkeit zu den elektronischen Komponenten bietet.

Nowadays audio transmission systems are an almost standard feature of guided tours in cities, factories or museums. However, they only permit communication in one direction, i.e. from the guide to the participants, precluding the possibility of questions. That is about to change with this newly developed audio guide, which is equipped with a »talkback« function that even allows questions to be put to the entire group. The function is activated by turning the slender device upside down, giving it the character of a handheld microphone.

Both its dimensions and soft contours are intended to ensure ease of handling for users of any age. The display showing the channel is behind a dark surface that transitions into the top of the device with the headphone jack. The housing is designed for easy replacement should damage occur and provides easy access to the electronic components.

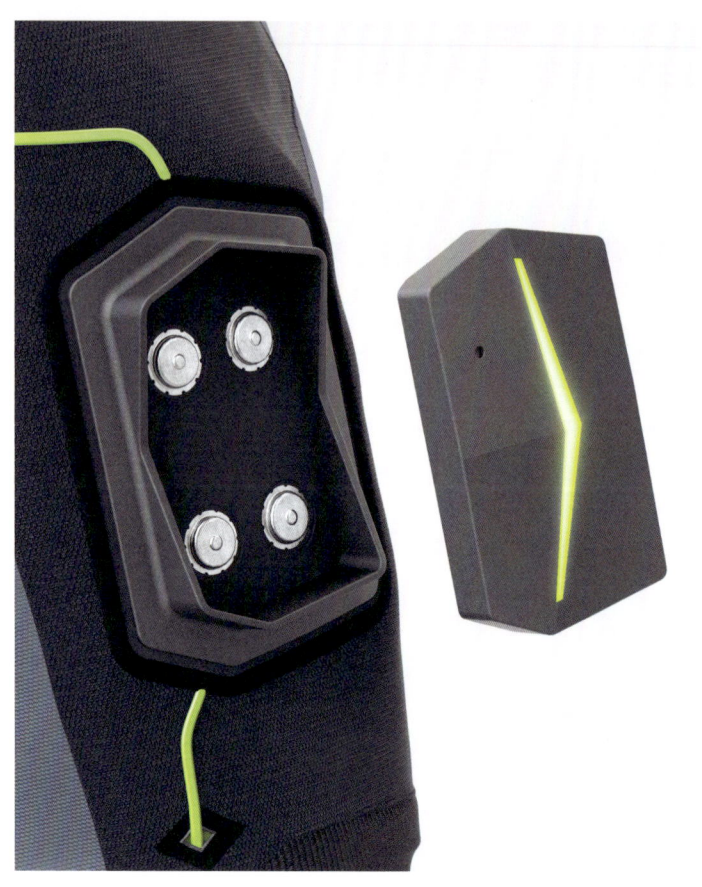

1 → SEITE/PAGE 186–191

2 → SEITE/PAGE 192, 193

MATERIALS & SURFACES

GOLD:
1 **ANGEL**
Adresys GmbH
Salzburg
Österreich/Austria

SPECIAL MENTION:
2 **AFILIA**
Rökona Textilwerke GmbH & Co KG
Tübingen

Technik beeinflusst schon immer das Design, auch Werkstoffe tun dies mehr denn je. Materialien mit innovativen Eigenschaften eröffnen Nutzungsszenarien, die zu mehr Nachhaltigkeit, geringerem Ressourcenverbrauch oder optimierter Funktionsintegration leiten.

Technology has always influenced design, and materials are having a greater impact than ever before. Materials with innovative properties open up usage scenarios that lead to greater sustainability, reduced resource consumption or the optimised integration of functions.

GOLD ANGEL SICHERHEITSSYSTEM
SAFETY SYSTEM

MATERIALS & SURFACES
MATERIALS & SURFACES

186
187

FOCUS
GOLD

JURY STATEMENT

Dieses Wearable kann Leben retten und ist uneingeschränkt zu begrüßen. Die Integration der Technik ist sehr gelungen und lässt die besondere Funktion des sportiv anmutenden T-Shirts zunächst nicht erkennen – sieht man von der zentralen Steuereinheit ab, die magnetisch am Ärmel fixiert wird.

This wearable can save lives and is unquestionably a very good thing. The integration of the technology is very compelling and does not reveal the sporty-looking T-shirt's special function at first glance – apart, that is, from the central control unit that is fixed to the sleeve magnetically.

HERSTELLER/MANUFACTURER
Adresys GmbH
Salzburg
Österreich/Austria

DESIGN
Yellow Design GmbH
Pforzheim

VERTRIEB/DISTRIBUTOR
Adresys GmbH
Salzburg
Österreich/Austria

Stromunfälle passieren häufig und nicht nur Elektroinstallateur:innen. Wichtig ist, dass bei solchen Vorfällen schnellstmöglich eine Rettung eingeleitet wird.

Das System Angel ist zwar keine Schutzausrüstung im klassischen Sinne, erkennt aber Notsituationen und kontaktiert automatisch Notrufzentralen. Angel verbindet Sensorik, Aktuatorik und Kommunikation in Form eines T-Shirts. Elektroden in den Bündchen erkennen Stromdurchgänge durch den Körper, andere Sensoren beobachten die Bewegung oder auch Reglosigkeit der Nutzer:innen. Die zentrale Steuereinheit am Ärmel, Genius genannt, verarbeitet die Sensordaten und weist über eine Bluetooth-Verbindung die zugehörige Smartphone-App an, den betroffenen Stromkreis sofort abzuschalten. Zugleich geht ein Notruf mit GPS-Daten an eine durchgehend aktive Zentrale. Der Task Timer wiederum alarmiert, wenn nach vordefinierter Zeitspanne keine manuelle Bestätigung erfolgt.

Electrical accidents happen often, and not just to electricians. When such incidents occur, it's crucial for the rescue operation to get underway as quickly as possible.

Although the Angel system is not safety equipment in the classic sense, it detects emergency situations and automatically alerts the first responders. Angel combines sensor, actuator and communication technology in the form of a T-shirt. Electrodes in the cuffs detect the passage of current through the body, other sensors monitor whether the user is moving or motionless. The central control unit on the sleeve – the Genius – processes the sensor data and, via Bluetooth, notifies the accompanying smartphone app to interrupt the affected circuit immediately. At the same time, an alert with GPS data is sent to a permanently active emergency call centre. Another function, the Task Timer, sounds the alarm if no manual acknowledgement is received after a predefined period of time has elapsed.

ALEXANDER SCHLAG GESCHÄFTSFÜHRER,
YELLOW DESIGN GMBH

»Wir durften eine ganz neue Produktkategorie vom weißen Blatt gestalten.«

»We got to design a totally new product category from scratch.«

ALEXANDER SCHLAG   **MANAGING PARTNER, YELLOW DESIGN GMBH**

Wann wurden Sie als Designagentur in den Entwicklungsprozess involviert?
Die Idee und die technische Umsetzung war derart neu, dass für die Entwicklung ein crossdisziplinäres und hoch spezialisiertes Team gebildet wurde. Der Charakter der Entwicklung war daher eher vergleichbar mit einem Start-up, da sämtliche Technologien und Prozesse neu entwickelt werden mussten – mit dem Unterschied, dass das Team bereits hoch professionell war.

Adresys wusste aus anderen Projekten, dass es zielführend ist, den Designer möglichst frühzeitig einzubinden, also waren wir bereits in der frühen Ideenphase dabei. Bei unserem Einstieg gab es allein die bahnbrechende Idee, mittels körpernaher Sensoren in einem Kleidungsstück, Stromunfälle zu erkennen und automatisiert eine Rettungskette in Gang zu setzen. Wie die Technologie in ein Wearable zu integrieren ist, war gänzlich unbekannt, entsprechende Technologien waren weltweit noch nicht verfügbar. Wir durften, was eine besonders schöne Aufgabe war, sämtliche wesentlichen Bausteine des Erscheinungsbildes und der Produktentwicklung vom weißen Blatt weg gestalten, dazu gehörten das Erscheinungsbild mit Wort- und Bildmarke und das Design für alle textilen und technischen Objekte.

Und wie tief waren Sie in die Entwicklung der Technologie einbezogen?
Die Technologie, also die elektronische Hard- und Software, ist Kernkompetenz von Adresys und wurde inhouse entwickelt. Die Herausforderungen des Projektes, also kleinste Bauformen, Verknüpfung von unterschiedlichsten Materialien und Technologien an den Oberflächen, setzte aber eine sehr enge Zusammenarbeit voraus. Es ging ja nicht darum, ein Design um ein Produkt zu hüllen, sondern insbesondere bei dem Shirt Technologie und Design im und am Textil selbst zu integrieren.

Welchen Umfang hatte Ihre Designleistung?
Wir haben das Projekt von der Idee über die Realisierung von Prototypen und umfangreichen Feldtests bis hin zur Serienreife begleitet.

Könnte die Technologie auch in anderen Kontexten hilfreich sein?
Ja, das ist bereits realisiert und damit auch für andere Berufszweige wirksam, die Sturzerkennung kann auch solitär arbeiten. Das bedeutet, dass auch bei anderen Unfallarten beziehungsweise anderen Auslösern, zum Beispiel Reglosigkeit aufgrund eines Sturzes von Leiter oder Gerüst die Rettungskette aktiviert wird. Noch weiter geht die neueste Erweiterung No-Motion. Sie ermöglicht auch die Erkennung von Ereignissen mit anschließender Bewegungs- oder Bewusstlosigkeit. Wird ein Voralarm nicht innerhalb einer definierten Zeit quittiert, dann wird ein Notfall erkannt und die Rettungskette aktiviert.

Die yellow design GmbH ist eine international tätige, multidisziplinäre Agentur für Designentwicklung in Pforzheim und Tokio. Vor rund 45 Jahren gegründet, ist sie das Fundament der yellow group, dem Agenturnetzwerk mit weiteren Standorten in Köln und Berlin. Seit 2011 ist Alexander Schlag geschäftsführender Gesellschafter des Unternehmens.

www.yellowdesign.com

As the design agency, when did you become involved in the development process?
The idea and technical implementation were so new that a cross-disciplinary and highly specialised team was formed for the development work. So in that respect the development was more like a startup in character, because all the technologies and processes had to be newly devised – but with the difference that the team was already highly professional.

Adresys knew from other projects that involving the designer as early on as possible is productive, so we were already on board when the idea was still in its infancy. When we joined the team, there was nothing but the groundbreaking idea of using sensors worn close to the body in a garment to detect electrical accidents and automatically set the emergency response chain in motion. As for how the technology could be integrated into a wearable, that was totally unknown, no corresponding technologies were available anywhere in the world. We were allowed to design all the essential components of the corporate design and the product development from scratch – it was a great assignment that included the wordmark and logo for the corporate design and the design of all the textile and technical objects.

And how deeply involved were you with the development of the technology?
The technology, i.e. the electronic hard- and software, is Adresys' core competency and was developed in-house. However, the challenges the project presented – things like tiny components, the combining of all sorts of different materials and technologies on the surfaces – called for very close collaboration. After all, it wasn't a question of wrapping a design around a product; instead, especially in the case of the shirt, we had to integrate the technology and design into and onto the textile itself.

What scope did your design work cover?
We accompanied the project from the idea to the realisation of prototypes and extensive field tests all the way to the production-ready stage.

Could the technology be helpful in other contexts too?
Yes, in fact the usage scenarios for the technology have already been extended to make it utilisable for other professions, and the fall detection can work on its own too. That means the emergency response chain can be activated in the case of different types of accident or causes as well, for instance if the wearer isn't moving after falling off a ladder or scaffolding. The latest version goes even further. It's called No Motion and permits the detection of events with subsequent motionlessness or unconsciousness. If a pre-alarm is not acknowledged within a defined period of time, the system recognises an emergency and activates the response chain.

yellow design GmbH is an internationally active, multidisciplinary agency for design development based in Pforzheim and Tokyo. Founded approx. 45 years ago, it is the foundation of the yellow group, an agency network that also has locations in Cologne and Berlin. Alexander Schlag has been managing partner at the company since 2011.

www.yellowdesign.com

SPECIAL MENTION | AFILIA
→ SEITE/PAGE 193

TEXTILES LEDER
TEXTILE LEATHER

SPECIAL MENTION — AFILIA — TEXTILES LEDER / TEXTILE LEATHER

JURY STATEMENT

Dieses textile Halbzeug weist ökologisch in die richtige Richtung – es ist sortenrein, frei von oft problematischen Additiven und nach Gebrauch recycelbar. Eine spannende Materialentwicklung, die nicht nur im automotiven Kontext Relevanz hat.

Ecologically speaking, this semi-finished product certainly points in the right direction: the textile is a mono-material, contains no additives – problematic or otherwise – and can be recycled after use. An intriguing materials development whose relevance is by no means limited to the automotive context.

HERSTELLER/MANUFACTURER
Rökona Textilwerke GmbH & Co KG
Tübingen

DESIGN
Inhouse

VERTRIEB/DISTRIBUTOR
Rökona Textilwerke GmbH & Co KG
Tübingen

Obwohl optisch einer Oberfläche aus Nubukleder gleich, besteht dieses Material nicht aus Leder, sondern zu 100 Prozent aus recyceltem Polyester – ist also eine vegane Alternative. Interessant dabei ist, dass der haptische und visuelle Effekt keine Zugabe spezieller Chemikalien erfordert, sondern durch die spezielle Adaption der Wirktechnologie entsteht. Das Textil ist überdies sortenrein und damit bestens wieder in den Stoffkreislauf rückführbar. Und im Gegensatz zu üblichen Verfahren kommt das Färbeverfahren ganz ohne Einsatz von Wasser aus.

Primär für die Automobilbranche entwickelt, also für Verkleidungen, Akzente oder Dachhimmel, steht der Verwendung als Heimtextil im Interior- oder im Modebereich nichts entgegen.

Although it might look like nubuck on the surface, this material is not made of leather at all. Instead, it consists entirely of recycled polyester – and is therefore a vegan alternative. It's interesting that the look and feel are achieved by adapting the warp knitting technology rather than by adding special chemicals. What's more, because the textile is a mono-material, its return to the loop is unproblematic. And unlike the processes commonly used, the dyeing method does not rely on water.

Mainly developed for the automotive sector, i.e. for trims, accents or headliners, nothing actually stands in the way of using the textile in home interiors or the fashion industry as well.

MIA SEEGER PRI[ZE]
2022
MIA SEEGER PRI[ZE]
2022

Jährlicher Wettbewerb der Mia Seeger Stiftung
für junge Designerinnen und Designer
mit Unterstützung der Hans Schwörer Stiftung und
des Rat für Formgebung

The Mia Seeger Foundation's annual
competition for young designers,
sponsored by the Hans Schwörer Foundation
and the German Design Council

MIA SEEGER PREIS 2022
MIA SEEGER PRIZE 2022

DIE JURY
THE JURY

IMMANUEL CHI
Industrial Designer/Designtheoretiker, Pforzheim; Mia Seeger Preisträger 1992
→ Industrial designer/design theoretician, Pforzheim; Mia Seeger prize winner 1992

JÜRGEN GEHM
Produktdesigner und Ökonom, Bosch Power Tools, Leinfelden-Echterdingen; Design Driven Innovation, Tübingen
→ Product designer and economist, Bosch Power Tools, Leinfelden-Echterdingen; Design Driven Innovation, Tübingen
www.juergengehm.de

BARBARA LERSCH
Kulturmanagerin,
Hans Sauer Stiftung, München
→ Cultural manager,
Hans Sauer Foundation, Munich

STEFAN LIPPERT
Designer, UP Designstudio, Stuttgart; Mia Seeger Preisträger 1993 und Stipendiat 1993/94
→ Designer, UP Designstudio, Stuttgart; Mia Seeger prize winner 1993 and scholarship winner 1993/94

DANIEL RAUH
Designer, Shift GmbH, Falkenberg (Wabern)
→ Designer, Shift GmbH, Falkenberg (Wabern)

OLIVER STOTZ
Industriedesigner, Wuppertal;
Mia Seeger Preisträger 1992
→ Industrial designer, Wuppertal;
Mia Seeger prize winner 1992
stotz-design.com

PROF.IN VERONIKA AUMANN
Textildesignerin, Staatliche Akademie der Bildenden Kunste Stuttgart
→ Textile designer, Stuttgart State Academy of Art and Design

*nicht im Bild/not pictured

MIA SEEGER PREIS 2022
MIA SEEGER PRIZE 2022

10.000 EURO FÜR JUNGE DESIGNERINNEN UND DESIGNER
€10,000 FOR YOUNG DESIGNERS

Zum 31. Mal vergibt die Mia Seeger Stiftung den Mia Seeger Preis über insgesamt 10.000 Euro an junge Designerinnen und Designer. Zum dritten Mal gehen die Mia Seeger Stiftung und die Bewerber dafür den digitalen Weg. Ein Spaziergang ist es noch nicht.

Die Stiftungs-Geschäftsführerin und die Mitarbeiterinnen im Design Center, Jurorinnen und Juroren legten sich ordentlich ins Zeug, damit aus den 108 Anmeldungen von 29 Hochschulen zum Schluss drei Preise und fünf Anerkennungen hervorgingen. An die eingereichten Arbeiten hatte die Jury neben den üblichen Qualitätsmaßstäben den des sozialen Nutzens anzulegen.

Dem Rat für Formgebung ist es zu danken, dass die Stiftung den Preis wieder in gewohnter Höhe ausschreiben konnte.

The Mia Seeger Foundation is presenting the Mia Seeger Prize to young designers for the 31st time, together with prize money totalling €10,000. And for the third time, circumstances have obliged the Mia Seeger Foundation and the contenders to take the digital route, which is not yet without its difficulties.

But that didn't deter the foundation's managing director, the Design Center team and the jurors from giving their all, and in the end the 108 entries received from 29 colleges and universities resulted in three prizes and five Highly Commended distinctions. In addition to the usual quality criteria, the jury also assessed the works submitted on the basis of their benefit to society.

Thanks to the German Design Council, the foundation was once again able to award the customary amount of prize money.

MIA SEEGER PREIS 2022
MIA SEEGER PRIZE 2022

JURY STATEMENT

Mit diesem Rollwagen passt das Haarewaschen endlich in den Pflegealltag, sowohl im Heim als auch zuhause. Gestalt und Struktur sind selbsterklärend; einer besonderen Schulung bedarf es nicht. »Hygiene ist ein Grundbedürfnis, trägt zum körperlichen Wohlbefinden bei und beschleunigt den Heilungsprozess.« Unter diesem Leitgedanken der Designerin ist die Aufgabe bestens gelöst und mit angemessenem Aufwand ein empfindlicher Mangel behoben. Bettlägerige wie Pflegekräfte erfahren dadurch Wertschätzung.

Thanks to this trolley, hair-washing is finally compatible with the day-to-day care routine, whether the patient is in a residential facility or their own home. Form and structure are self-explanatory, no special training is required. »Hygiene is a basic need, contributes to physical wellbeing and speeds up the healing process.« With this guiding principle as her starting point, the designer has come up with an excellent solution and eliminated a serious deficit in the care-giving routine with appropriate means. Both the bedridden and their carers feel more valued as a result.

ENTWURF/DEVELOPER
Helena Kiefer
helena.kiefer@t-online.de

STUDIUM/DEGREE COURSE
Industrie-Design Diplom
Hochschule Darmstadt/
University of Applied Sciences

BETREUUNG/SUPERVISOR
Prof. Tom Philipps

Bestehende Hilfsmittel für die Haarwäsche im Liegen in Krankenhäusern und Pflegeheimen sind häufig unhandlich und zeitraubend. Das ausgezeichnete Diplom-Projekt der Studentin Helena Kiefer widmet sich diesem Problem des pflegerischen Alltags und zeichnet sich durch einen sensiblen Rechercheprozess aus. Mit den von Immobilität Betroffenen und potentiellen Nutzerinnen und Nutzern wird sehr ausführlich gesprochen.

Vaask – ein System zur Haarwäsche im Liegen – löst das Problem. Der Rollwagen enthält Frisch- und Gebrauchtwassertanks und hat Stauraum für Zubehör. Obenauf das Waschbecken. Es ist an einer Seite mit einem weichen Silikoneinsatz für eine ergonomische Nackenauflage versehen. Mit angeschlossenem Ablaufschlauch wird es am Kopfende im Bett platziert. Das Haarewaschen geht dann wie beim Frisör. Besonderes Augenmerk wurde auf die Form- und Farbgestaltung gelegt, die im klinischen Umfeld stimmungsaufhellend wirken.

Existing aids for washing the hair of hospital patients and care home residents who are confined to their beds are often awkward to handle and time-consuming. The prize-winning finals project by student Helena Kiefer addresses this day-to-day care-giving problem and stands out for its sensitive research approach, which involves in-depth interviews with both the immobile bed occupants and the potential users.

Vaask – a system for washing the hair of care recipients in a recumbent position – solves the problem. The trolley contains tanks for fresh and used water and provides storage space for equipment. The washbasin sits on top and features a soft silicone inset on one side that provides ergonomic support for the neck. Positioned at the head end of the bed with the drainage hose attached, the basin thus allows the hair to be washed just like at a salon. The design pays special attention to form and colour, both of which have an uplifting effect in their clinical setting.

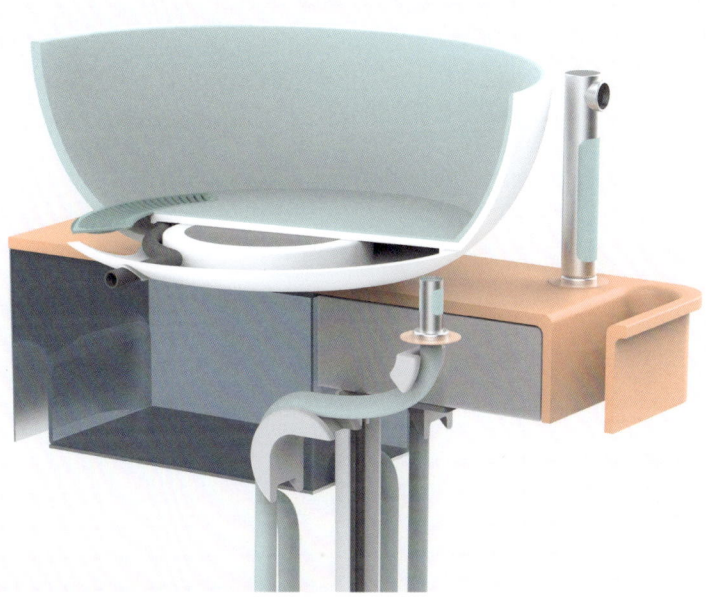
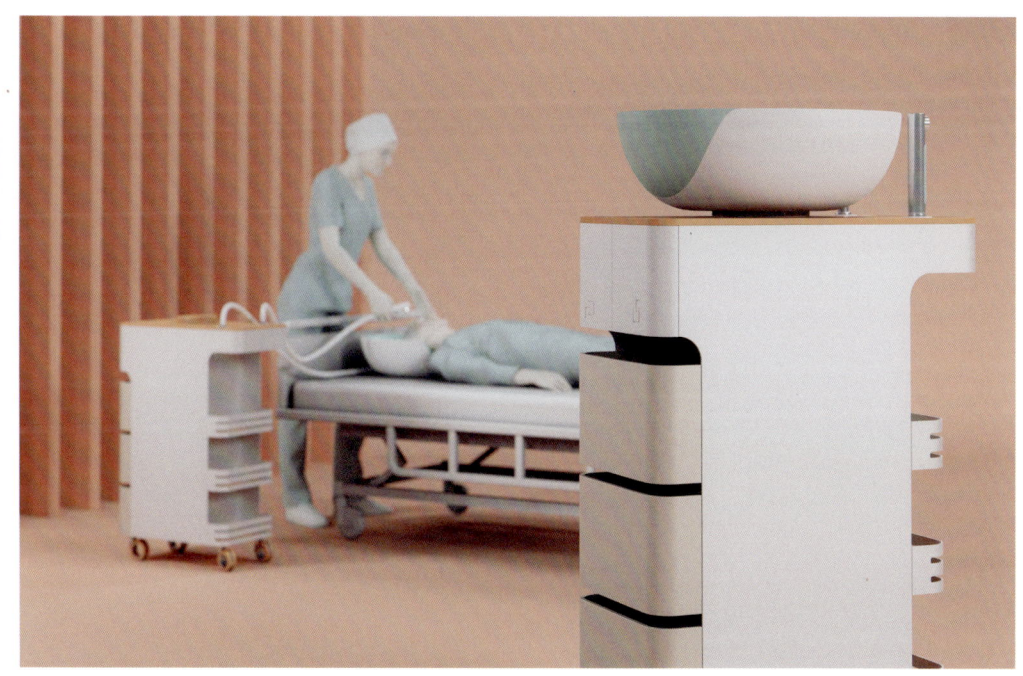

MIA SEEGER PREIS 2022
MIA SEEGER PRIZE 2022

JURY STATEMENT

Allgemein formulieren die Autorinnen der vorgelegten Studie ihren Anspruch so: »Wie Gestalter eine differenzierte Meinungsbildung in Zeiten eines zunehmenden Populismus unterstützen können.« In einem Punkt aber werden sie konkreter: Es gilt, populistische Nachrichten mit grafischen Populismus-Markern zu versehen. Der Gegner soll nicht mit den eigenen Waffen, sondern mit seriöseren Mitteln aus dem visuellen Fach geschlagen werden. Das kommt an gut ausgewählten Bildbeispielen sowie im Entwurf einer Kampagne gegen das Dagegensein zum Tragen und bestätigt sich in Interviews mit interessanten Partnerinnen und Partnern. Hierin und in der theoretischen Vorarbeit hat die Arbeit ihre Stärken.

The authors of the study formulate their general intent as follows: »How designers can support nuanced opinion forming in times of increasing populism.« However, there is one point where they become more concrete: they propose using graphic populism markers to draw attention to populist messages. The aim is to defeat adversaries not with their own weapons but with more reputable means from the visual toolbox. This approach is applied not just to well-chosen sample images but to the design of an »anti-being-anti« campaign, and is borne out in interviews conducted with interesting partners. It is these elaborations and the theoretical groundwork that constitute the project's strengths.

Die ausgezeichnete Arbeit »The New Anti« von Natalie Kohler und Annika Tessmer ist eine Aufklärungsinitiative gegen Populismus. Eine Analyse von typischen Merkmalen populistischer Botschaften in den Medien steht an ihrem Anfang. Dann sucht sie nach den Möglichkeiten, solche Merkmale visuell, also mit grafischen Mitteln, zu markieren und damit die manipulierende Agitation zu entlarven. Aus einzelnen Beispielen, die auf Instagram kommuniziert sind, entsteht eine visuelle Strategie, die erst zu einem Wissenskatalog verdichtet, dann in einem Web-Seminar verbreitet und schließlich in einem Workshop vermittelt wird. Die Initiative richtet sich vorwiegend an junge Menschen.

Entitled The New Anti, the prize-winning project by Natalie Kohler and Annika Tessmer is an educational initiative that aims to combat populism. It begins by analysing the typical characteristics of populist messages in the media and then looks for ways to indicate such characteristics visually, i.e. using graphic markers, with the goal of exposing their manipulative attempts to agitate. Individual examples, shared on Instagram, are used as the basis for a visual strategy that is first condensed into a knowledge catalogue, then disseminated in a web seminar and finally communicated in a workshop. The initiative is primarily aimed at young people.

ENTWURF/DEVELOPERS
Annika Tessmer
annika.tessmer@gmail.com
Natalie Kohler
natalie_kohler@gmx.de

STUDIUM/DEGREE COURSE
Strategische Gestaltung MA
Hochschule für Gestaltung
Schwäbisch Gmünd

BETREUUNG/SUPERVISORS
Prof. Dr. Susanne Schade
Prof. Carmen Hartmann-Menzel

THE NEW ANTI
THE NEW ANTI

MIA SEEGER PREIS 2022
MIA SEEGER PRIZE 2022

JURY STATEMENT

Mit Blick auf die Not in fernen Ländern hat der Designer ein Gerät entwickelt, das in Gestalt seiner eigenen Bauanleitung überall im Eigenbau entstehen kann. Alles dazu Nötige und hier schon Mögliche ist bedacht und vorbereitet. Vor Ort brauchen die Beteiligten freilich noch Initiative, Wissenszugang, Lerneifer, Findigkeit und nicht zuletzt, im regionalen Kontext, viel Geld. Darum sind auch die Kosten bedacht. Glaubhaft belegen sie den Anspruch, mit dem Entwurf zum Abbau der weltweiten Chancenungleichheit beitragen zu können, durchaus im Sinne der »Sustainable Development Goals« der Vereinten Nationen.

Mindful of the poverty in faraway countries, the designer has developed a device and instructions for making it, allowing the machine to be built anywhere. Everything that's necessary and can be done remotely has been considered and prepared. Obviously those involved at local level still need initiative, access to knowledge, a desire to learn, resourcefulness and what, at least in a regional context, is a lot of money. That's why the costs have been taken into account as well. They are credible proof of the design's aspiration to contribute to the elimination of global inequality – an approach very much in keeping with the United Nations' sustainable development goals.

ENTWURF/DEVELOPER
Elias Grieninger
elias.grieninger@gmail.com

STUDIUM/DEGREE COURSE
Produktgestaltung BA
Hochschule für Gestaltung
Schwäbisch Gmünd

BETREUUNG/SUPERVISOR
Diplom-Designer Stefan Lippert

Ein Dialysegerät ist teuer. Weltweit haben 80 % von über 10 Mio. Menschen, die eine Dialyse-Behandlung brauchen, keinen Zugang zu einer Behandlung. Mit einem mechanischen Grundmodul aus dem 3D Drucker, auf dem standardisierte Filter- und Blutschlauchsysteme montiert werden, wird es eher möglich.

Der Entwicklung des Projekts liegt der Open Source Gedanke zu Grunde. Der Designer stellt die Pläne, Druckdateien, Steuerungsprogramme und alles notwendige technische Know-How als Hilfe zur Selbsthilfe der Welt zur Verfügung. Per Internet sind die Daten verfügbar, so dass vor Ort mit den vorhandenen Mitteln ein Gerät entstehen kann.

Die einzelnen Bauteile inklusive des speziellen Blut-Filters des von Elias Grieninger konzipierten Gerätes sind als Standard-Bauteile weltweit für unter 100 Dollar erhältlich, die Software ist kostenlos, Drucker- und Materialkosten sind überschaubar. Was könnte jetzt noch einer weltweiten Verbreitung im Wege stehen?

A dialysis machine is expensive. Of the more than 10 million people who need dialysis worldwide, 80% have no access to treatment. A mechanical basic model made by a 3D printer and equipped with standardised filter and blood tubing systems could change that for the better.

The project's development is based on the open source idea. In an effort to help people help themselves, the designer makes the plans, print files, control programmes and all the necessary technical know-how available to the world. The data is accessible online, allowing a device to be made anywhere with the means available.

The individual components, including the special blood filter used in the machine designed by Elias Grieninger, are standard parts that can be purchased all over the world for under $100; the software is free, the costs for printing and materials moderate. It would seem as if there's nothing standing in the way of worldwide distribution.

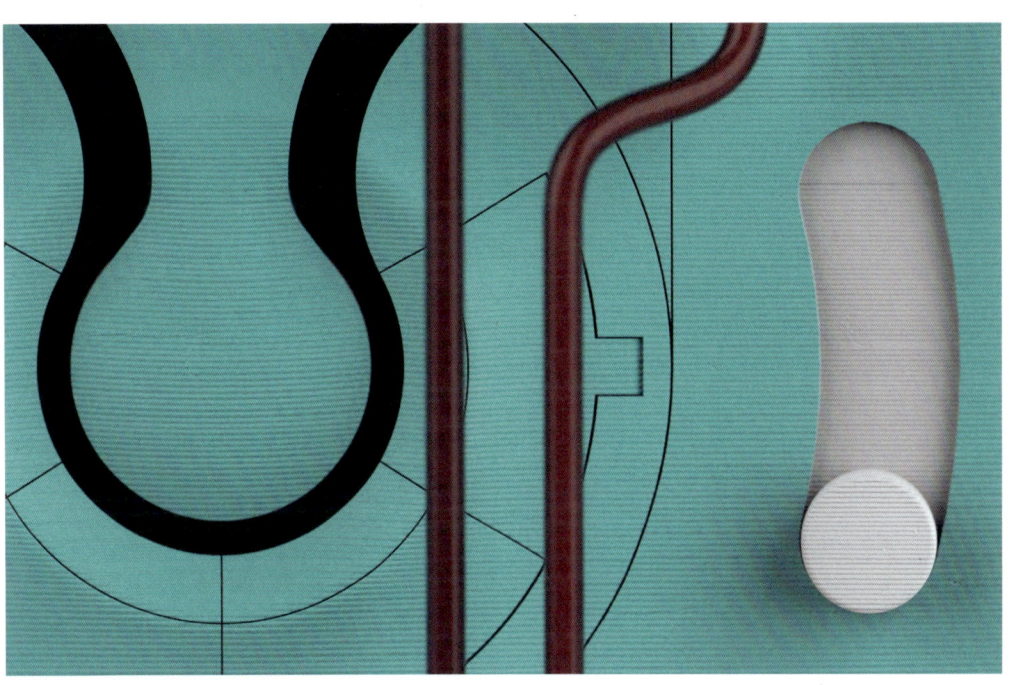
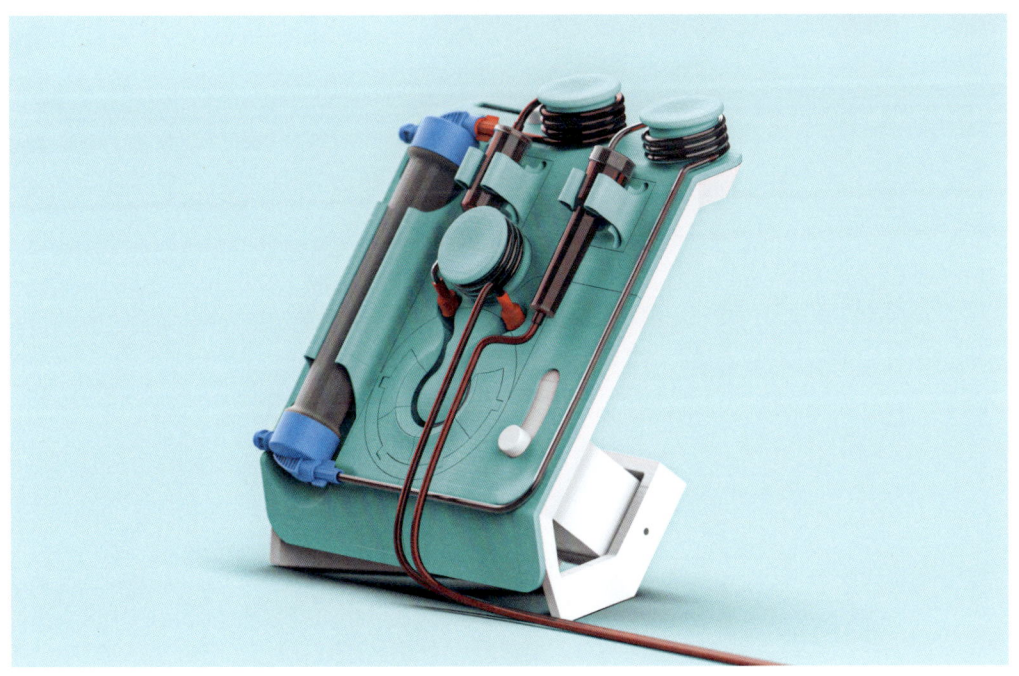

MIA SEEGER PREIS 2022　　DER MIA SEEGER STIFTUNG
MIA SEEGER PRIZE 2022　　PRESENTED BY THE MIA SEEGER FOUNDATION

WAS MEHR ALS EINEM NÜTZT

Auch in diesem Jahr wurden wieder Entwürfe von Produkten ausgezeichnet, die sich mit wichtigen Aspekten unseres Lebens und Zusammenlebens befassen und hierfür neuartige, sinnvolle Lösungen vorschlagen. Dabei soll der Art, wie Menschen – beruflich oder privat, alt oder jung, gesund oder krank – untereinander kommunizieren und miteinander umgehen, besonderes Augenmerk gelten.

BENEFITING MORE THAN THE INDIVIDUAL

This year too, the prizes and commendations were presented to product designs that address important aspects of our lives and living together and propose innovative, meaningful solutions. Particular attention was to be focused on how people communicate and interact with one another – in a professional or private setting, young or old, healthy or ill.

ANERKENNUNG
HIGHLY COMMENDED

MONKEYBOTS
MONKEYBOTS

JURY STATEMENT

Eine Technologie, die in der Industrie vielfach und erfolgreich Anwendung findet, öffnet sich auf herzerfrischende Weise der privaten Nutzung. Wer sich damit auskennt und die genannten Gerätschaften hat oder beides sich aneignet, kann recht unkompliziert kleine Helferlein oder Kobolde für Haus und Hof selber bauen. Schnittstellen zu anderen Anwendungen sind mitgedacht.

A technology that is already finding successful and widespread application in industry is now opening up to private contexts in a thoroughly refreshing way. Anybody who has or is willing to acquire the necessary skills and devices will have no trouble building their own little elves and helpers for use and fun around the home. Interfaces with other applications have been factored in.

ENTWURF/DEVELOPERS
Georg Kloeck
gk@kloeckwork.com
Mohammad Moradi
hi@momoradi.com

STUDIUM/DEGREE COURSE
Produkt-Design/Interaction Design
Weißensee Kunsthochschule Berlin

BETREUUNG/SUPERVISOR
Prof. Carola Zwick

MonkeyBots kommen aus dem Internet der Dinge, kurz IoT. In einen kleinen Würfel ist ein Motor und digitale Technologie integriert. Smartphone mit App machen ihn zum Roboter, der sich von dort aus programmieren und steuern lässt. Mit verschiedenen Aufsätzen, die aus dem 3D Drucker kommen, kann er andere Dinge drehen, schütteln, drücken oder kippen. Ernsthafte und lustige Vorschläge zur Anwendung im persönlichen Umfeld von Erwachsenen und Kindern erläutern das Konzept.

MonkeyBots come from the internet of things, or the IoT for short. A motor and digital technology are integrated into a small cube. Using a smartphone and app, it can be turned into a robot, which is then programmed and controlled in the same way. Equipped with various tools made by a 3D printer, the MonkeyBot can rotate, shake, push or tilt other things. The concept is explained in the form of serious and funny suggestions for potential applications in the personal environments of adults and children.

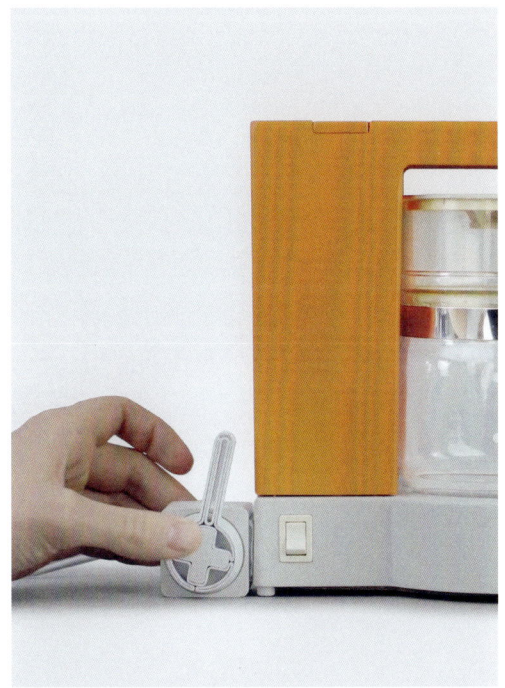 

ANERKENNUNG
HIGHLY COMMENDED

EYETALK – KOMMUNIZIEREN IM LOCKED-IN SYNDROM
EYETALK – COMMUNICATING IN LOCKED-IN SYNDROME

JURY STATEMENT

Verständigung durch ein winziges Nadelöhr. In umfangreichen Recherchen im Umfeld betroffener Patienten haben Designerin und Designer doch noch eine Lösung gefunden. Mit der Entscheidung für die Elektrookulografie blieb ihnen eine aufwendige Erfassung von Augenbewegungen mit Kamera und anschließender Bilderkennung erspart. Der Rest war klassische, sorgfältig ausgeführte Entwurfsarbeit, hin zu einer formal perfekten Kombination aus Headset und Tischgerät.

Communication through the tiny eye of a needle. And yet thanks to their extensive research in direct proximity with sufferers, the designers have nevertheless managed to find a solution. By deciding to use electrooculography, they saved themselves the complex process of capturing eye movements with a camera and tracking them with image recognition software. The rest was classic, meticulously executed development work, resulting in a perfectly designed combination of headset and tabletop device.

Ein Locked-in Syndrom hat erlitten, wer nichts mehr bewegen kann, meist nur noch die Augen. Andere Signale zum Kommunizieren hat er nicht mehr. EyeTalk registriert sie über Elektroden als Spannungsdifferenzen und meldet sich über Bone Conduction Kopfhörer beim Patienten mit einer Satzerstellungssoftware. Im Hin und Her von sprachlichen Vorschlägen und Augenbewegungen entsteht der beabsichtigte Satz. Für alle Beteiligten hörbar, ertönt er zum Schluss aus der Dockingstation.

People suffering from locked-in syndrome are unable to move any part of their body; the eyes are usually the only exception. Eye movements are therefore the only way they can send signals to communicate with the outside world. Via electrodes, eyeTalk registers these movements as potential differences and uses bone conduction headphones to give the patient access to sentence-generating software. The back-and-forth between suggestions for linguistic output and eye movements results in the intended sentence, which is ultimately voiced by the docking station so that it's audible for everybody.

ENTWURF/DEVELOPERS
Jonas Mahler
Jonas.mahler@hfg.design
Amelie Straubmüller

STUDIUM/DEGREE COURSE
Produktgestaltung BA
Hochschule für Gestaltung
Schwäbisch Gmünd

BETREUUNG/SUPERVISORS
Prof. Gerhard Reichert
Andreas Hess
Gastbetreuer Karl-Eugen Siegel

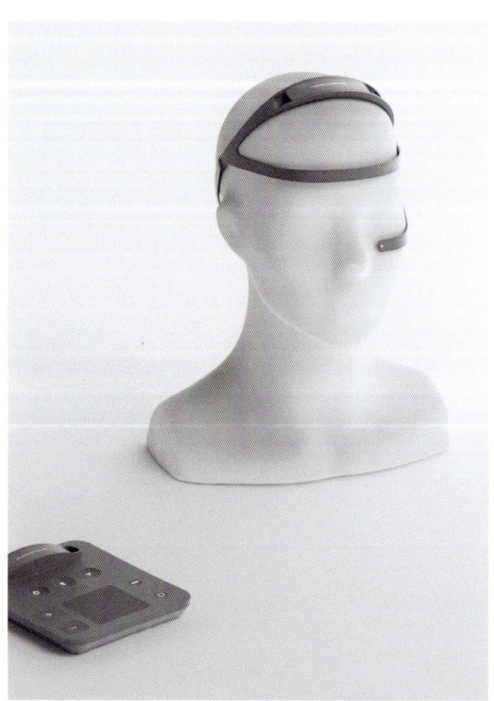
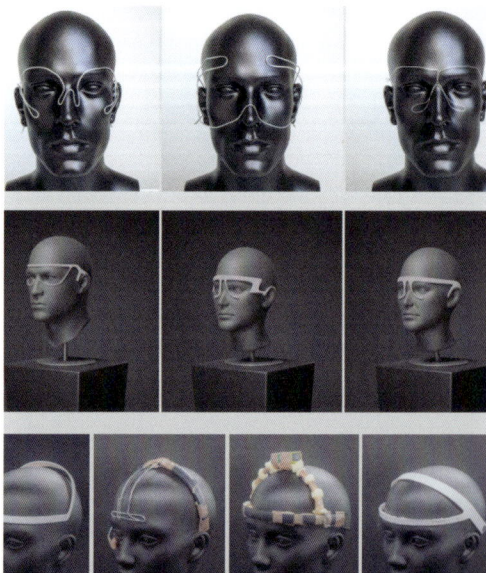

ANERKENNUNG
HIGHLY COMMENDED

REPAIRABLE BY DESIGN
REPAIRABLE BY DESIGN

JURY STATEMENT

Der Nutzen von Reparierbarkeit liegt auf der Hand: Produkte langlebig machen, Ressourcen schonen, Schrott vermeiden.

Ein wichtiges Thema ist aufgegriffen, interessant bearbeitet und mit den Karten zu brauchbaren Handreichungen verdichtet. Zu beachten ist, dass nicht nur dem Design die Verantwortung in der Sache der Reparierbarkeit zufällt, die ganze Kette aus Marketing, Design, Entwicklung und Herstellung trägt dazu bei. Wobei es allen zugutekommt, wenn im Entwurf schon vorbedacht ist, was im Produkt perfektioniert wird.

The benefit of repairability is obvious: it increases product longevity, conserves resources and avoids scrap.

An important issue has been addressed, presented in an interesting way and condensed into cards that provide useful guidance. It's important to note that design doesn't bear sole responsibility for repairability: the entire chain of marketing, design, development and production has a role to play. At the same time, everybody benefits if the concept already anticipates what's ultimately perfected in the product.

ENTWURF/DEVELOPERS
David Schöllhorn
david.schoellhorn@icloud.com
Vincent Propst
vincentpropst@gmx.de

STUDIUM/DEGREE COURSE
Produktgestaltung BA
Hochschule für Gestaltung
Schwäbisch Gmünd

BETREUUNG/SUPERVISORS
Prof. Gabriele N. Reichert
Prof. Matthias Held

Früher war mehr Reparatur. Die Industrie verkauft lieber Neuware. Allmählich regt sich Widerstand. Verbraucherinnen, Verbraucher und Umweltschutz rufen nach dem Gesetzgeber. Doch wie will man Reparierbarkeit definieren? Dazu sind hier vom Produktdesign her Vorschläge erarbeitet, in zehn Richtlinien zusammengefasst. Drei Entwürfe – Ohrhörer, Kaffeeautomat und elektrische Zahnbürste – zeigen, wie die Anwendung der Regeln aussehen können.

Repairs used to be more common. Nowadays, however, industry prefers to sell new goods. But resistance is gradually growing. Consumers and environmentalists are calling for legislation. But how do you go about defining repairability? Condensed into 10 guidelines, the suggestions presented here take product design as their starting point. Three designs – earbuds, a coffee maker and an electric toothbrush – show what application of the rules could look like in practice.

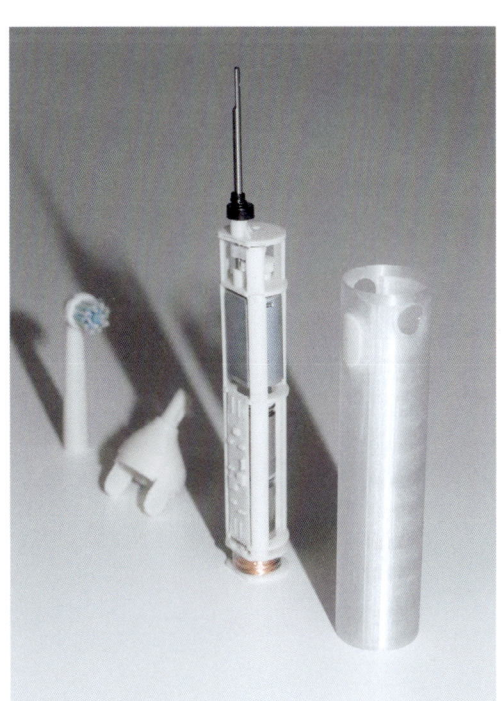
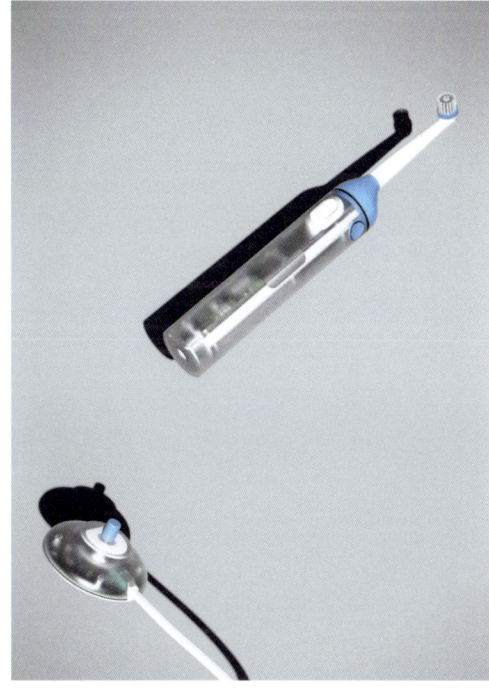

ANERKENNUNG
HIGHLY COMMENDED

PAPER PREGNANCY TEST
PAPER PREGNANCY TEST

JURY STATEMENT

Ein millionenfach gekauftes Produkt wird einer gründlichen Revision unterzogen und nochmals neu gedacht. Es geht auch ohne Plastik! Das ist die Botschaft dieses Entwurfs. Und gilt überall dort, wo nur die Befestigung für einen Teststreifen und eine Verpackung gebraucht wird. In diesem Punkt weist der Entwurf über das Nachweisen einer Schwangerschaft weit hinaus. Die Hersteller von Corona-Tests möchten bitte einen Blick drauf werfen.

A product that sells in its millions has been revisited and thoroughly revised. We can manage without plastic! That's the message behind this design – and it can be transferred to any product that essentially only consists of a holder and packing for a test strip. In that respect, the design's potential goes far beyond pregnancy tests. The makers of corona tests should take a good look at it as well.

Eine Untersuchung gängiger Schwangerschaftstests ergab: Ihr Kernstück ist der Teststreifen, und der ist aus Papier; drumherum, bei Halterung und Verpackung wird mit Plastik nicht gegeizt. Versuche zeigen aber: Recyceltes Papier tut's auch, nämlich Pappmaschee für die Streifenhalterung, Karton für die Verpackung. Das Testergebnis lässt sich an zwei kleinen Fenstern mit unmissverständlicher Markierung ablesen. Nach Gebrauch gehört alles zusammen in den kompostierbaren Müll.

A study of bestselling pregnancy tests revealed that the core component is the test strip – which is made of paper. The holder and packaging that surround it, on the other hand, are made of copious amounts of plastic. However, experiments demonstrated that recycled paper is perfectly adequate: papier-mache for holding the test strip, card for the packaging. The result is revealed via two little windows with unambiguous markings. After use, the whole thing can be disposed of as compostable waste.

ENTWURF/DEVELOPER
Yue Zhao
zzzy0604@gmail.oom

STUDIUM/DEGREE COURSE
Industriedesign MA
Muthesius Kunsthochschule Kiel

BETREUUNG/SUPERVISOR
Prof. Detlef Rhein

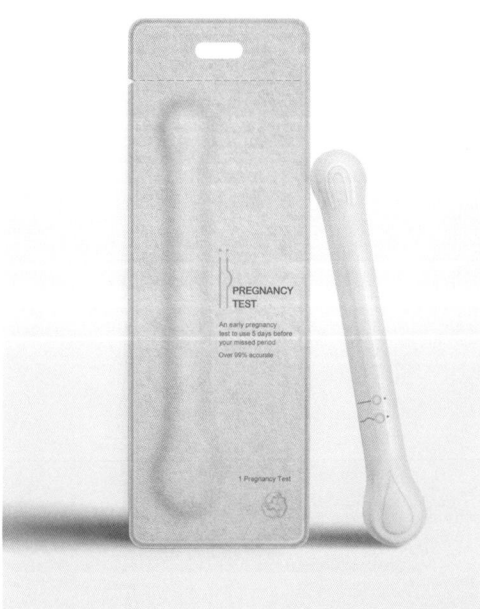

ANERKENNUNG
HIGHLY COMMENDED

THEATER ON TOUR
THEATER ON TOUR

JURY STATEMENT

Der Entwurf eines fliegenden Gebäudes stellt sich in den Dienst einer sozialen Aufgabe. Als abgewandelter Prototyp gebaut, wird es am jeweiligen Ort für eine Woche zum Kristallisationspunkt für vorbereitete Feste, organisierte Treffen und zwanglose Begegnungen lokaler Institutionen und freier Initiativen, natürlich auch zur Spielstätte für einstudierte Stücke. Es entsteht ein temporärer Ort für Austausch, Begegnung und Diskurs.

Das Projekt überwindet die Grenzen der klassischen Produktgestaltung und ist der wagemutige Versuch, durch ein Objekt eine Intervention in einem nicht gestaltungsaffinen Kontext darzustellen und damit einen Diskurs zu initiieren. Die »Gute Form« des Objektes ist die Interface-Funktion, die das Objekt zweifellos erfüllt.

The design of a temporary building aims to perform a social function. Constructed as a modified prototype, it spends a week at its respective location, where it serves local institutions and independent initiatives as a focal point for prearranged festivities, organised meetings and informal gatherings, and of course as a stage for rehearsed performances. It creates a temporary space for exchange, encounter and discourse.

The project transcends the boundaries of classic product design and is a bold attempt to use an object as a means of intervention in a non-design-savvy context and thus initiate a discourse. The »good design« of the object lies in the interface function that it unquestionably performs.

ENTWURF/DEVELOPER
Ezra Dilger
ezra-dilger@gmx.de

STUDIUM/DEGREE COURSE
Industrial Design MA
Burg Giebichenstein
Kunsthochschule Halle/
University of Art and Design

BETREUUNG/SUPERVISOR
Prof. Guido Englich

Eine Tribüne mit vorgelagerter Bühne, transportabel und einfach auf- und wieder abzubauen, ist Teil eines realisierten Theaterprojekts, das sich der kulturellen Entwicklung »im ländlichen Raum« verschrieben hat. Das »Kollektiv Plus X e.V.« zieht von Kleinstadt zu Kleinstadt, nimmt Kontakt zu soziokulturellen Einrichtungen vor Ort auf und initiiert in Kooperation mit ihnen Workshops und Aufführungen. Die Leute, vorwiegend junge, die zur Landflucht neigen, sollen ihre Interessen zur Geltung bringen können.

Audience seating with a stage in front of it, transportable and easy to erect and dismantle, is part of an already implemented theatre project dedicated to cultural development »in the rural space«. The Kollektiv Plus X e.V. moves from small town to small town, makes contact with local socio-cultural facilities and collaborates with them to initiate workshops and performances. The project aims to enable people who might be inclined to relocate to cities – mainly youngsters – to raise awareness of their interests.

MIA SEEGER STIFTUNG

THE MIA SEEGER FOUNDATION

IMPRESSUM/PUBLISHING DETAILS

HERAUSGEBER/PUBLISHED BY
Mia Seeger Stiftung

REDAKTION/EDITOR
Marion Ascherl, Schwäbisch Gmünd

ÜBERSETZUNG/TRANSLATION
Alison Du Bovis, Jork

GRAFIKDESIGN/GRAPHIC DESIGN
Julian Hölzer,
stapelberg & fritz, Stuttgart

AUSSTELLUNGSGESTALTUNG/ EXHIBITION DESIGN
Thomas Simianer, Stuttgart

FOTOS/PHOTOS
Preisträger, Ausgezeichnete/
Prize winners, commended entrants,
Karl Fisch

KOORDINATION MIT »FOCUS OPEN«/ COORDINATION WITH FOCUS OPEN
Hildegard Hild

JURYVORBEREITUNG/ JUDGING ORGANISED BY
Renate Seeger, Iris Steinmetz,
Marion Ascherl,
Team UP Designstudio

DIGITALE TECHNIK, VIDEO-KONFERENZEN/ DIGITAL TECHNOLOGY, VIDEO CONFERENCES
Stefan Lippert, UP Designstudio

Mia Seeger Stiftung
c/o Design Center
Baden-Württemberg
im Haus der Wirtschaft
Willi-Bleicher-Straße 19
D-70174 Stuttgart
T +49 711 123 2781
F +49 711 123 2771

E-Mail: design@rps.bwl.de
www.mia-seeger.de
instagram.com/miaseeger

Abbildung rechts: Mia Seeger in der Zentrale des Deutschen Werkbundes in Berlin, 1928; Foto: Cami Stone, Stadtarchiv Stuttgart aus dem Nachlass Mia Seeger/
Right: Mia Seeger at the headquarters of the Deutscher Werkbund in Berlin, 1928; photo: Cami Stone, from the Mia Seeger papers held by Stuttgart City Archives

Mia Seeger war die »Grande Dame« des Design. Mit der Weißenhofsiedlung 1927 in Stuttgart begann ihre Laufbahn. Bald war sie an weiteren Ausstellungen des Deutschen Werkbundes beteiligt.

Die Bundesrepublik hat sie vielfach als Kommissarin zu Triennalen in Mailand entsandt und zur ersten Leiterin des Rat für Formgebung berufen, den sie zwölf Jahre lang führte. Sie war selbst keine Designerin, sondern Design-Vermittlerin und -Beraterin. 1986 rief sie die nach ihr benannte Stiftung ins Leben, deren Zweck die Bildung junger Gestalterinnen und Gestalter ist. Namhafte Sponsoren aus der Wirtschaft haben sich ihren Zielen angeschlossen.

Mit der Absicht, besonders den Nachwuchs im Design zu fördern und ihn dabei zur Auseinandersetzung mit sozialen Fragen aufzufordern, schreibt die Stiftung jährlich den Mia Seeger Preis unter dem Motto »was mehr als einem nützt« aus. Seit Jahren kann sie die Ergebnisse ihres Designwettbewerbs im Rahmen der Ausstellung »Focus Open – Internationaler Designpreis BadenWürttemberg« präsentieren. Dafür ist sie dem Design Center sehr dankbar, auch für technische und organisatorische Vorbereitungen der Jurierung.

Darüber hinaus erfreut sich die Stiftung schon länger einer vertraglich vereinbarten Kooperation mit dem Rat für Formgebung. Er trägt auch dazu bei, das Wirken der Stiftung, insbesondere die Kontinuität des Mia Seeger Preises finanziell zu sichern. Auch andere Spenderinnen und Spender haben mit einmaligen Zuwendungen dazu beigetragen, 2016 Alexander Neumeister und im Jahr danach die Hans-Schwörer-Stiftung. Wer in dieser oder ähnlicher Weise die gemeinnützige Arbeit der Mia Seeger Stiftung unterstützen möchte, wendet sich am besten an die Geschäftsführerin der Stiftung, Marion Ascherl.

Über ihre Arbeit informiert die Stiftung auf ihrer Internetseite: www.mia-seeger.de
Darüber hinaus gibt es News und Posts rund um Design mit sozialem Anspruch auf: instagram.com/miaseeger

Mia Seeger was the »grande dame« of design. Her career began with the Weissenhof Estate in Stuttgart in 1927. She was soon involved with further exhibitions by the Deutscher Werkbund as well.

The Federal Republic of Germany sent her to the Triennial exhibitions in Milan as its commissioner multiple times and appointed her the first director of the German Design Council, which she headed for 12 years. She herself was not a designer but a design mediator and adviser. She established the foundation that bears her name in 1986 for the purpose of promoting young designers' education. Renowned sponsors from commerce and industry have joined the foundation in the pursuit of its goals.

With the specific aim of promoting young designers and challenging them to tackle social issues, the foundation invites entries for the annual Mia Seeger Prize under the motto »benefiting more than the individual«. For some years now, it has been able to present the results of its design competition within the context of the Focus Open – Baden-Württemberg International Design Award exhibition. The foundation is deeply obliged to the Design Center for its assistance, as well as for the technical and organisational support it provides for the judging process.

In addition, the foundation has had a cooperation agreement with the German Design Council for some time now. This contract helps ensure the financial security of the foundation's work, and in particular the continuity of the Mia Seeger Prize. Other benefactors have also provided valuable support in the form of one-off donations, including Alexander Neumeister in 2016 and the Hans Schwörer Foundation in the following year. Anybody who would like to support the Mia Seeger Foundation's non-profit work in this or a similar way should please contact the foundation's managing director Marion Ascherl

Detailed information about the foundation's work is available on its website: www.mia-seeger.de
The foundation also publishes news and posts about design with a social slant at Instagram.com/miaseeger.

APPENDIX
A—Z

ADRESSEN/ ADDRESSES

A

ADRESYS GMBH
Oberndorferstraße 35
A-5020 Salzburg
T +43 59495 6900
www.adresys.com
S/P 189

AGILOX SERVICES GMBH
Gewerbepark 3
A-4671 Neukirchen bei Lambach
T +43 7245 93083 0
www.agilox.net
S/P 25

AME B.V.
Esp 100
NL-5633 AA Eindhoven
T +31 40 264 6400
www.ame.nu
S/P 171

ARCHÄOLOGISCHES LANDESMUSEUM BADEN-WÜRTTEMBERG
Benediktinerplatz 5
78467 Konstanz
T +49 7531 9804 0
www.alm-bw.de
S/P 146
Simon Neßler

B

BENE GMBH
Schwarzwiesenstraße 3
A-3340 Waidhofen an der Ybbs
T +43 7442 500
www.bene.com
S/P 85
Christian Horner

BESSEY TOOL GMBH & CO. KG
Mühlwiesenstraße 40
74321 Bietigheim-Bissingen
T +49 7142 4010
www.bessey.de
S/P 41

BLEICHERT AUTOMATION GMBH & CO. KG
Hans-Ulrich-Breymann-Straße 35
74706 Osterburken
T +49 6291 93 0
www.bleichert.de
S/P 42

BOTTA DESIGN
Klosterstraße 15a
61462 Königstein/Taunus
T +6174 96 11 88
www.botta-design.de
S/P 95

C

COUCOU GMBH
Mathildenstraße 15a
45130 Essen
T +49 171 8748 313
www.coucou.group
S/P 111

D

DEFORTEC GMBH
Breitwasenring 15
72135 Dettenhausen
T +49 7157 7211 820
www.defortec.de
S/P 43

DQBD GMBH
Schulstraße 15
73614 Schorndorf
T +49 7181 937 666 0
www.dqbd.de
S/P 128

E

ERCO GMBH
Brockhauser Weg 80-82
58507 Lüdenscheid
T +49 2351 551 100
www.erco.com
S/P 103

F

FISCHERWERKE GMBH & CO. KG
Klaus-Fischer-Straße 1
72178 Waldachtal
T +49 7443 12 0
www.fischer.de
S/P 40

FISCHER DEUTSCHLAND VERTRIEBS GMBH
Klaus-Fischer-Straße 1
72178 Waldachtal
T +49 7443 12 6000
www.fischer.de
S/P 40

FISCHER SPORTS GMBH
Fischerstraße 8
A-4910 Ried im Innkreis
T +43 7752 909 329
www.fischersports.com
S/P 119

FORMQUADRAT GMBH
Brucknerstraße 3-5
A-4020 Linz
T +43 732 777 244
www.formquadrat.com
S/P 19, 25, 119

G

GBO INNOVATION MAKERS
Wethouder den Oudenstraat 6
NL-5706 ST Helmond
T +31 88 888 69 00
www.gbo.eu
S/P 171

GROHE DEUTSCHLAND VERTRIEBS GMBH
Zur Porta 9
32457 Porta Westfalica
T +49 571 3989 333
www.grohe.de
S/P 57

GROHE AG
Feldmühleplatz 15
40545 Düsseldorf
T +49 211 9130 3000
www.grohe.com
S/P 57

GROSS + FROELICH GMBH & CO. KG
Josef-Beyerle-Straße 9
71263 Weil der Stadt
T +49 7033 522 3
www.gross-stabil.de
S/P 87

H

HAUS DER GESCHICHTE BADEN-WÜRTTEMBERG
Konrad-Adenauer-Straße 16
70173 Stuttgart
T +40 711 212 39 50
www.hdgbw.de
S/P 147

I

ID AID GMBH
Vogelsangstraße 12
70176 Stuttgart
T +49 711 27 350 888
www.idaid.com
S/P 137

IMAGE CONSTRUCTION MESSE- UND EVENTBAU GMBH
Luxemburger Straße 7
41812 Erkelenz
T +49 2431 805 150
www.image-construction.com
S/P 111

INSTAGRID GMBH
Hermann-Hagenmeyer-Straße 1
71636 Ludwigsburg
T +49 7141 6962 430
www.instagrid.co
S/P 31
Felix Fuchs

J

JEHS & LAUB
Römerstraße 51A
70180 Stuttgart
T +49 711 620 7390
www.jehs-laub.com
S/P 84

K

KID SYSTEMS GMBH
Lüneburger Schanze 30
21614 Buxtehude
T +49 40 743 71633
www.kid.systeme.de
S/P 45

KIMETEC GMBH
Gerlinger Straße 36–38
71254 Ditzingen
T +49 7156 17602 0
www.kimetec.com
S/P 51

KLANGERFINDER GMBH & CO. KG
Humboldtstraße 4
70178 Stuttgart
T +49 711 656 799 88
www.klangerfinder.de
S/P 147

KUCKOO CAMPER GMBH & CO. KG
Industriestraße 20
74369 Löchgau
T +49 7143 39 415 0
www.kuckoo-camper.de
S/P 170

L

BURKHARDT LEITNER MODULAR SPACES GMBH
Olgastraße 138
70180 Stuttgart
T +49 711 255 880
www.burkhardtleitner-units.com
S/P 149
Akin Nalca

LIXIL GLOBAL DESIGN
Feldmühleplatz 15
40545 Düsseldorf
T +49 211 9130 3216
www.grohe.com
S/P 57

M

MAXMAIER URBANDEVELOPMENT
Schwieberdinger Straße 72
71636 Ludwigsburg
T +49 7141 479 0
www.urbanharbor.com
S/P 155

MEDER COMMTECH GMBH
Robert-Bosch-Straße 4
78224 Singen
T +49 7731 911 322 0
www.meder-commtech.de
S/P 183

MIDDELHAUVEDESIGN
Auf dem Holleter 26
45138 Essen
T +49 172 2133 225
www.middelhauve.net
S/P 111

MIELE & CIE. KG
Carl-Miele-Straße 29
33332 Gütersloh
T +49 5241 89 0
www.miele.de
S/P 63, 70, 71

MOBIMEX AG
Birren 17
CH-5703 Seon
T +41 62 769 7000
www.studiobymobimex.com
S/P 84

N

NABERTHERM GMBH
Bahnhofstraße 20
28865 Lilienthal
T +49 4598 922 0
www.nabertherm.com/de
S/P 179

NIMBUS GROUP GMBH
Sieglestraße 41
70469 Stuttgart
T +49 711 63 30 14 0
www.nimbus-lighting.com
S/P 137

O

OTTENWÄLDER UND OTTENWÄLDER
Sebaldplatz 6
73525 Schwäbisch Gmünd
T +49 7171 927230
www.ottenwaelder.de
S/P 42

P

PICA-MARKER GMBH
Sonnengarten 11
91356 Kirchehrenbach
T +49 9191 320 403 0
www.pica-marker.com
S/P 94
Stephan Möck

PRIMO GMBH
Wernher-von-Braun-Straße 2
84544 Aschau am Inn
+49 8638 88559 200
www.primo-gmbh.com
S/P 136

PURE POSITION BY IWL GGMBH
Traubinger Straße 23
82346 Machtlfing
T +49 8157 931 416
www.pureposition.de
S/P 86

R

RAUMHOCHN GESTALTUNG IM RAUM GBR
Sauerweg 2
70563 Stuttgart
T +49 179 768 1451
www.raumhochn.de
S/P 148

RECARO AIRCRAFT SEATING GMBH & CO. KG
Daimlerstraße 21
74523 Schwäbisch Hall
T +49 791 503 7000
www.recaro-as.com
S/P 165

RED GMBH
Wilhelmstraße 5a
70182 Stuttgart
T +49 711 248 521 01
www.redproducts.de
S/P 72

RÖKONA TEXTILWERKE GMBH & CO. KG
Schaffhausenstraße 101
72072 Tübingen
T +49 7071 153 0
www.roekona.de
S/P 193

ROSENBAUER INTERNATIONAL AG
Paschinger Straße 90
A-4060 Leonding
T +43 732 6794 0
www.rosenbauer.com
S/P 19

ROSENBAUER DEUTSCHLAND GMBH
Rudolf-Breitscheid-Straße 79
14943 Luckenwalde
T +49 3371 6905 0
www.rosenbauer.com
S/P 19

R3
Wilhelmstraße 5a
70182 Stuttgart
T +49 711 248 521 01
S/P 72

S

GEORG SCHLEGEL GMBH & CO. KG
Kapellenweg 4
88525 Dürmentingen
T +49 7371 5020
www.schlegel.biz
S/P 44

**ANDREAS SCHULZE/
INDUSTRIAL DESIGN**
Cahensylstraße 3b
65549 Limburg an der Lahn
T +49 6431 971 6961
www.schulze-design.de
S/P 136

**OLAF SCHROEDER
INDUSTRIAL DESIGN**
Stubenrauchstraße 58
12161 Berlin
T +49 30 3060 2706
www.olafschroeder.com
S/P 86

SERIEN RAUMLEUCHTEN GMBH
Hainhäuser Straße 3-7
63110 Rodgau
T +49 6106 69090
www.serien.com
S/P 107
Jean-Marc da Costa

SFP ARCHITEKTEN GMBH
Möhringer Straße 60/1
70199 Stuttgart
T +49 711 47 687 0
www.sfp-architekten.de
S/P 155

SOLIDFLUID PRODUKTDESIGN
Turmstraße 8
78467 Konstanz
T +49 7531 9450 230
www.solidfluid.de
S/P 183

SPEEDFAB GMBH
Schulstraße 15
73614 Schorndorf
T +49 7181 991 2992
www.speedfab.eu
S/P 128

**STAATLICHES MUSEUM
FÜR NATURKUNDE**
Rosenstein 1
70191 Stuttgart
T +49 711 8936 0
www.naturkundemuseum-bw.de
S/P 148

STOFFBOOT
Bernhard-Göring-Straße 96
04275 Leipzig
T +49 174 35 65 804
www.stoffboot.de
S/P 126
Ringo Köhler

SUPERNOVA DESIGN GMBH
Industriestraße 26
79194 Gundelfingen
T +49 761 600 629 0
www.supernova-lights.com
S/P 129

T

TEAMS DESIGN GMBH
Kollwitzstraße 1
73728 Esslingen
T +49 711 351 765 0
www.teamsdesign.com
S/P 40

TRIPLE A MARKETING GMBH
Am Lenkwerk 3
33609 Bielefeld
T +49 521 1639 5972
S/P 127

U

UP DESIGNSTUDIO GMBH & CO. KG
Dornierstraße 17
70469 Stuttgart
T +49 711 3265460
www.updesignstudio.com
S/P 51

V

J.M. VOITH SE & CO. KG
Voith Paper
Escher-Wyss-Straße 25
88212 Ravensburg
T +49 7321 37 0
www.voith.com
S/P 43

W

WD3 GMBH
Seidenstraße 57
70174 Stuttgart
T +49 711 284 977 20
www.wd3.design
S/P 88

WEINBERG & RUF GBR
Martinstraße 5
70794 Filderstadt
T +49 711 7085 010
www.weinberg-ruf.de
S/P 41

WINKELBAUER-DESIGN
Myliusstraße 3
71638 Ludwigsburg
T +49 7141 903 222
www.winkelbauer-design.de
S/P 94

Y

YELLOW DESIGN GMBH
Bissingerstraße 6
75172 Pforzheim
T +49 7231 457 640
www.yellowdesign.com
S/P 189

Z

ZWEIGRAD GMBH & CO. KG
Donnerstraße 20
22763 Hamburg
T +49 40 3290 4748 0
www.zweigrad.de
S/P 45, 179

NAMENSREGISTER/ INDEX OF NAMES

A

ADRESYS GMBH
S/P 189
AGILOX SERVICES GMBH
S/P 25
AME B.V.
S/P 171
ARCHÄOLOGISCHES LANDESMUSEUM BADEN-WÜRTTEMBERG
S/P 146

B

BENE GMBH
S/P 85
BESSEY TOOL GMBH & CO. KG
S/P 41
BLEICHERT AUTOMATION GMBH & CO. KG
S/P 42
BOTTA DESIGN
S/P 95

C

COUCOU GMBH
S/P 111

D

DA COSTA, JEAN-MARC
S/P 107
DEFORTEC GMBH
S/P 43
DQBD GMBH
S/P 128

E

ERCO GMBH
S/P 103

F

FISCHERWERKE GMBH & CO. KG
S/P 40
FISCHER DEUTSCHLAND VERTRIEBS GMBH
S/P 40
FISCHER SPORTS GMBH
S/P 119
FORMQUADRAT GMBH
S/P 19, 25, 119
FUCHS, FELIX
S/P 31

G

GBO INNOVATION MAKERS
S/P 171
GROHE DEUTSCHLAND VERTRIEBS GMBH
S/P 57
GROHE AG
S/P 57
GROSS + FROELICH GMBH & CO. KG
S/P 87

H

HAUS DER GESCHICHTE BADEN-WÜRTTEMBERG
S/P 147
HORNER, CHRISTIAN
S/P 85

I

ID AID GMBH
S/P 137
IMAGE CONSTRUCTION MESSE- UND EVENTBAU GMBH
S/P 111
INSTAGRID GMBH
S/P 31

J

JEHS & LAUB
S/P 84

K

KID SYSTEMS GMBH
S/P 45
KIMETEC GMBH
S/P 51
KLANGERFINDER GMBH & CO. KG
S/P 147
KÖHLER, RINGO
S/P 126
KUCKOO CAMPER GMBH & CO. KG
S/P 170

L

BURKHARDT LEITNER MODULAR SPACES GMBH
S/P 149
LIXIL GLOBAL DESIGN
S/P 57

M

MAXMAIER URBANDEVELOPMENT
S/P 155
MEDER COMMTECH GMBH
S/P 183
MIDDELHAUVEDESIGN
S/P 111
MIELE & CIE. KG
S/P 63, 70, 71
MOBIMEX AG
S/P 84
MÖCK, STEPHAN
S/P 94

N

NABERTHERM GMBH
S/P 179
NALCA, AKIN
S/P 149
NESSLER, SIMON
S/P 146
NIMBUS GROUP GMBH
S/P 137

O

OTTENWÄLDER UND OTTENWÄLDER
S/P 42

P

PICA-MARKER GMBH
S/P 94
PRIMO GMBH
S/P 136
PURE POSITION BY IWL GGMBH
S/P 86

R

RAUMHOCHN
S/P 148
**RECARO AIRCRAFT SEATING
GMBH & CO. KG**
S/P 165
RED GMBH
S/P 72
**RÖKONA TEXTILWERKE
GMBH & CO. KG**
S/P 193
**ROSENBAUER
DEUTSCHLAND GMBH**
S/P 19
**ROSENBAUER
INTERNATIONAL AG**
S/P 19
R3
S/P 72

S

GEORG SCHLEGEL GMBH & CO. KG
S/P 44
**ANDREAS SCHULZE/
INDUSTRIAL DESIGN**
S/P 136
**OLAF SCHROEDER
INDUSTRIAL DESIGN**
S/P 86
SERIEN RAUMLEUCHTEN GMBH
S/P 107
SFP ARCHITEKTEN GMBH
S/P 155
SOLIDFLUID PRODUKTDESIGN
S/P 183
SPEEDFAB GMBH
S/P 128
**STAATLICHES MUSEUM
FÜR NATURKUNDE**
S/P 148
STOFFBOOT
S/P 126
SUPERNOVA DESIGN GMBH
S/P 129

T

TEAMS DESIGN GMBH
S/P 40
TRIPLE A MARKETING GMBH
S/P 127

U

**UP DESIGNSTUDIO
GMBH & CO. KG**
S/P 51

V

**J.M. VOITH SE & CO. KG
VOITH PAPER**
S/P 43

W

WD3 GMBH
S/P 88
WEINBERG & RUF GBR
S/P 41
WINKELBAUER-DESIGN
S/P 94

Y

YELLOW DESIGN GMBH
S/P 189

Z

ZWEIGRAD GMBH & CO. KG
S/P 45, 179

LET'S THANK …

GRAFIKDESIGN
GRAPHIC DESIGN
stapelberg&fritz gmbh
Julian Hölzer
Daniel Fritz

AUSSCHREIBUNG
CALL FOR ENTRIES

ANMELDUNG
REGISTRATION

TEAM FOCUS OPEN
Hildegard Hild
Michael Kern
Iris Steinmetz

TEXT & REDAKTION
TEXT & EDITORIAL SUPERVISION
Armin Scharf

JURY
Roland de Fries
Dina Gallo
Joa Herrenknecht
Andreas Hess
Marc-Gregor Weidt
Irmy Wilms-Haverkamp

JURIERUNG
JUDGING

JAHRBUCH
YEARBOOK

LEKTORAT
COPY-EDITING
Dr. Petra Kiedaisch
Gabriele Betz
Armin Scharf

VERLAG & VERTRIEB
PUBLISHING & DISTRIBUTION
avedition
Dr. Petra Kiedaisch

ÜBERSETZUNG
TRANSLATION
Alison Du Bovis

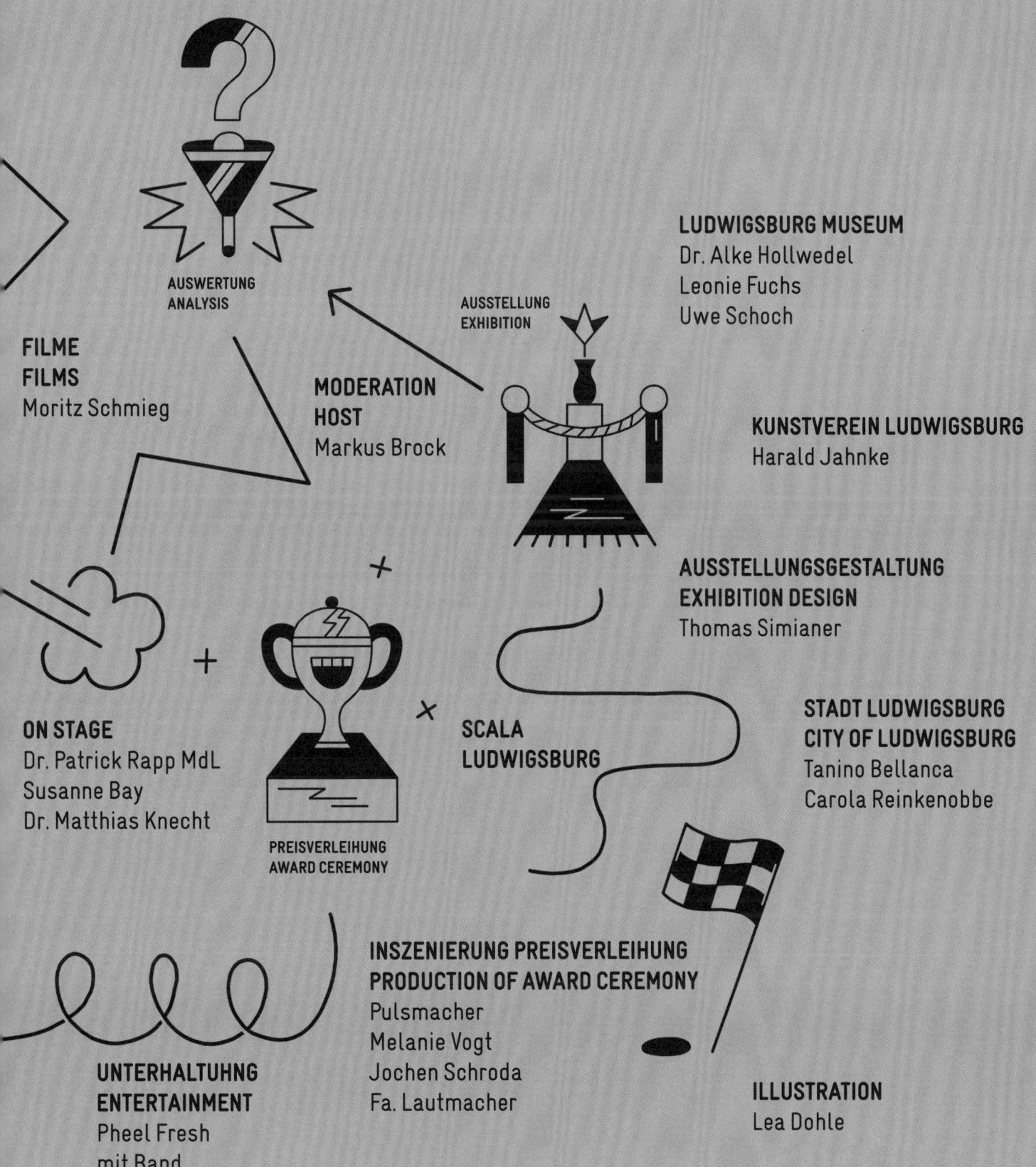

DESIGN IM DIALOG

Beratung, Fortbildung, Information und Präsentationen – das Design Center Baden-Württemberg ist eine nicht-kommerzielle Plattform für Design-Profis, Einsteiger und Unternehmer zugleich

DESIGN LESE
Vorträge, Medienpräsentationen und Diskussionsrunden zu aktuellen Themenbereichen aus Industrie, Design, Technik, Forschung und Wirtschaft.

DESIGN LESE LECTURES
Lectures, media presentations and panel discussions on up-to-the-minute topics from industry, design, technology, research and business.

EINSICHTEN
Austauschplattform für Industrie, Designwirtschaft, Forschung und Ausbildung. Unternehmen, Designagenturen und auch Design-Ausbildungsstätten erhalten die Möglichkeit, sich im Haus der Wirtschaft in Stuttgart detailliert zu präsentieren.

EINSICHTEN PRESENTATION PLATFORM
A platform for industry, the design sector, research and education where companies, design agencies and design schools are given the opportunity to stage detailed presentations at the Haus der Wirtschaft in Stuttgart.

DESIGN1ST BERATUNG
Im Rahmen unserer kostenfreien Design1st Beratung erhalten Unternehmer*innen Auskunft zu allen Fragen rund um Designleistungen und zu direkten Kooperationsmöglichkeiten mit der Designwirtschaft.

DESIGN1ST ADVISORY SERVICE
Our free Design1st advisory service provides entrepreneurs with information about anything to do with design services and advises them on the possibilities for direct cooperation with the design sector.

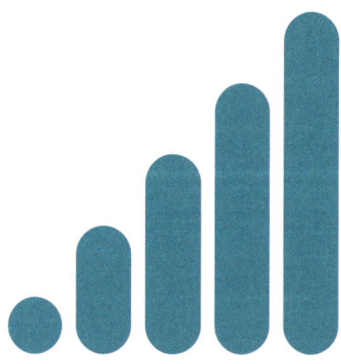

FIT FOR MARKET
Der richtige Schutz innovativer Produkte, die Anmeldung von Marken, die Honorierung kreativer Leistung oder die Vertragsgestaltung mit Designer*innen sind Themenfelder dieser Veranstaltungsreihe.

FIT FOR MAKET
This series of events covers topics like the right protection for innovative products, registering trademarks, appropriate payment for creative services and contractual arrangements with designers.

DESIGN IN DIALOGUE

Advice, training, information and presentations – the Design Center Baden-Württemberg is a non-commercial platform aimed not just at design professionals but at newcomers and entrepreneurs too.

DESIGN CENTER ROADSHOW
Veranstaltungen mit und bei unterschiedlichsten externen Kooperationspartnern, als Foren des Austauschs zwischen Industrie und Designwirtschaft.

DESIGN CENTER ROADSHOW
Events hosted by a wide range of external cooperation partners as forums where industry and the design sector can swap ideas and views.

DESIGN BIBLIOTHEK
Präsenzbibliothek für Designprofis und Designinteressierte, mit Online-Katalog und einem spezialisierten Publikationsbestand von rund 10.000 Büchern rund um das Thema Gestaltung.

DESIGN LIBRARY
A bricks-and-mortar library for design professionals and anyone interested in design, with an online catalogue and a specialised collection of around 10,000 publications on all aspects of design.

ENTDECKT
Die Präsentationsplattform für den Designnachwuchs! Vielversprechende Designtalente erhalten die Möglichkeit, sich samt ihrer aktuellen Projekte im Design Center der breiten Öffentlichkeit zu präsentieren.

ENTDECKT SHOWCASE
A presentation platform for up-and-coming designers that gives promising and talented newcomers the chance to introduce themselves and their latest projects to a broad public at the Design Center.

KONGRESSE & WORKSHOPS
Veranstaltungen zur Vermittlung von Know-how aus den unterschiedlichsten designrelevanten Disziplinen und Forschungsbereichen, aber auch aus dem weiten Feld des Marketings.

CONGRESSES & WORKSHOPS
Events that share know-how from all sorts of design-relevant disciplines and research areas, as well as from the broad field of marketing.

IMPRESSUM/ PUBLISHING DETAILS

HERAUSGEBER/PUBLISHER
Design Center Baden-Württemberg
Regierungspräsidium Stuttgart
Willi-Bleicher-Straße 19
70174 Stuttgart
T +49 711 123 26 84
design@rps.bwl.de
www.design-center.de

**TEXT UND REDAKTION/
TEXT AND EDITORIAL SUPERVISION**
Armin Scharf
Tübingen
www.bueroscharf.de

LEKTORAT/COPY-EDITING
Petra Kiedaisch
Armin Scharf
Gabriele Betz
Tübingen
www.gabriele-betz.de

ÜBERSETZUNG/TRANSLATION
Alison Du Bovis
Jork
www.dubovis.de

GRAFIKDESIGN/GRAPHIC DESIGN
stapelberg&fritz GmbH
Julian Hölzer
Stuttgart
www.stapelbergundfritz.com

**FOTOS DER JURY/
PHOTOS OF THE JURY**
Thomas Simianer

ILLUSTRATIONEN/ILLUSTRATIONS
Lea Dohle Illustration Stuttgart
www.leadohle.de
Instagram @leadohle

LITHOGRAFIE/LITHOGRAPHY
Corinna Rieber Prepress
www.rieber-prepress.de

DRUCK/PRINTING
Offizin Scheufele GmbH & Co. KG
Stuttgart
www.scheufele.de

PAPIER/PAPER
Juwel Offset,
PEFC-zertifiziert/
PEFC certified

**VERLAG UND VERTRIEB/
PUBLISHING AND DISTRIBUTION**
av edition GmbH
Senefelderstraße 109
70176 Stuttgart
T +49 711/2202279-0
kontakt@avedition.de
www.avedition.de

© 2022
av edition GmbH,
Design Center Baden-Württemberg
und die Autoren/and the authors

Alle Rechte vorbehalten./
All rights reserved.

ISBN 978-3-89986-382-6
Printed in Germany

Die Publikation erscheint
anlässlich der Ausstellung
»Focus Open 2022 –
Internationaler Designpreis
Baden-Württemberg
und Mia Seeger Preis 2022«

15. Oktober
bis 20. November 2022

This catalogue is published to
accompany the exhibition
»Focus Open 2022 –
Baden-Württemberg International
Design Award and
Mia Seeger Prize 2022«

15 October
to 20 November 2022

VERANSTALTER/ORGANISER
Design Center Baden-Württemberg
Regierungspräsidium Stuttgart
Willi-Bleicher-Straße 19
70174 Stuttgart
T +49 711 123 26 84

**VERANTWORTUNG UND KONZEPTION/
RESPONSIBILITY AND CONCEPT**
Christiane Nicolaus

**PROJEKTLEITUNG/
PROJECT MANAGER**
Hildegard Hild

ORGANISATION/ADMINISTRATION
Michael Kern

**AUSSTELLUNGSGESTALTUNG/
EXHIBITION DESIGN**
Thomas Simianer

**INSZENIERUNG PREISVERLEIHUNG/
PRODUCTION OF AWARD CEREMONY**
pulsmacher
Ludwigsburg
www.pulsmacher.de